D1433649

SOUTHERN AFRICA

Spectacular world of wildlife

SOUTHERN AFRICA

Spectacular world of wildlife

The Reader's Digest Association South Africa (Pty) Ltd, Cape Town

Principal Consultant
Professor John Skinner, Director, Mammal Research Institute,
University of Pretoria

Text: Alan Duggan, Peter Joyce

Editor: Tim O'Hagan
Art Editor: Christabel Hardacre
Project Co-ordinator: Carol Adams
Research: Judy Beyer, Mariëlle Renssen

SOUTHERN AFRICA, SPECTACULAR WORLD OF WILDLIFE
was edited and designed by
The Reader's Digest Association South Africa (Pty) Ltd,
130 Strand Street, Cape Town 8001

ISBN 1-874912-05-X

FACING TITLE PAGE: *A female cheetah and her cub at rest in the savannah.*

OPPOSITE: *A young nyala scans its woodland home for signs of danger.*

Front row seats in the theatre of the wild

Nature's masterpiece on elephant skin: a rare portrait of a wrinkled pachyderm.

There's nothing quite like it anywhere in the world.

It's a visual tour de force *– complete with high drama, raw tragedy, sound effects and stunning costumes. Scripted by Nature and choreographed by the weather, hunger and sheer chance, it's the spectacular world of southern African wildlife.*

The variety of our landscape, the extremes of climate, and the huge diversity of our vegetation create ideal conditions for a fraternity of wild animals unequalled anywhere else on earth. More than 900 species of birds cohabit freely with 338 mammal species, a diverse population of reptiles and amphibians, and an inestimable number of insects.

In South Africa alone, nearly six million hectares of land are devoted to conservation. There is a growing awareness of the importance of our wildlife heritage, and the necessity for its continued survival, in spite of industrial development and the encroachment of man.

Fuelling this awareness is the commitment of a group of dedicated photographers who spend days, months and years in the wild pursuing a dream: to capture on film the extraordinary lives of animals in a way that's never been done before.

The Spectacular World of Wildlife *presents more than 350 outstanding photographs taken by southern Africa's leading wildlife photographers in pursuit of this dream. The photographs uniquely reflect the savage wonder, the beauty and the fascination of the community of animals in their natural state.*

From bat-eared foxes, which use their ears like metal detectors to scan subterranean passages for signs of life, to chacma baboons that rock their babies in their arms, this book covers a wide and fascinating spectrum of animal behaviour. The savagery of a lion kill; the balletic leaps of a pronking springbok; the tender affection between an elephant and its calf ... it's all here, in a wildlife volume you'll never forget.

Poetry in motion: a herd of Burchell's zebras flees across grassy woodland in the great spaces of Botswana.

Contents

Siesta for a scavenger: a spotted hyaena cub takes a break in the shade of a bush.

The distribution of wildlife in southern Africa depends largely on the physical nature of a particular environment. Lions, for example, are at home in the bushveld-savannah and Kalahari regions, while hippos are found in the waterways. For this reason, the book is organised by natural region – as shown on the map opposite and in the contents list below.

Supplementing the regional sections are special 'portfolio' features covering important aspects of animal behaviour. The book ends with 'Where to see wildlife', detailing the major game reserves of southern and east Africa.

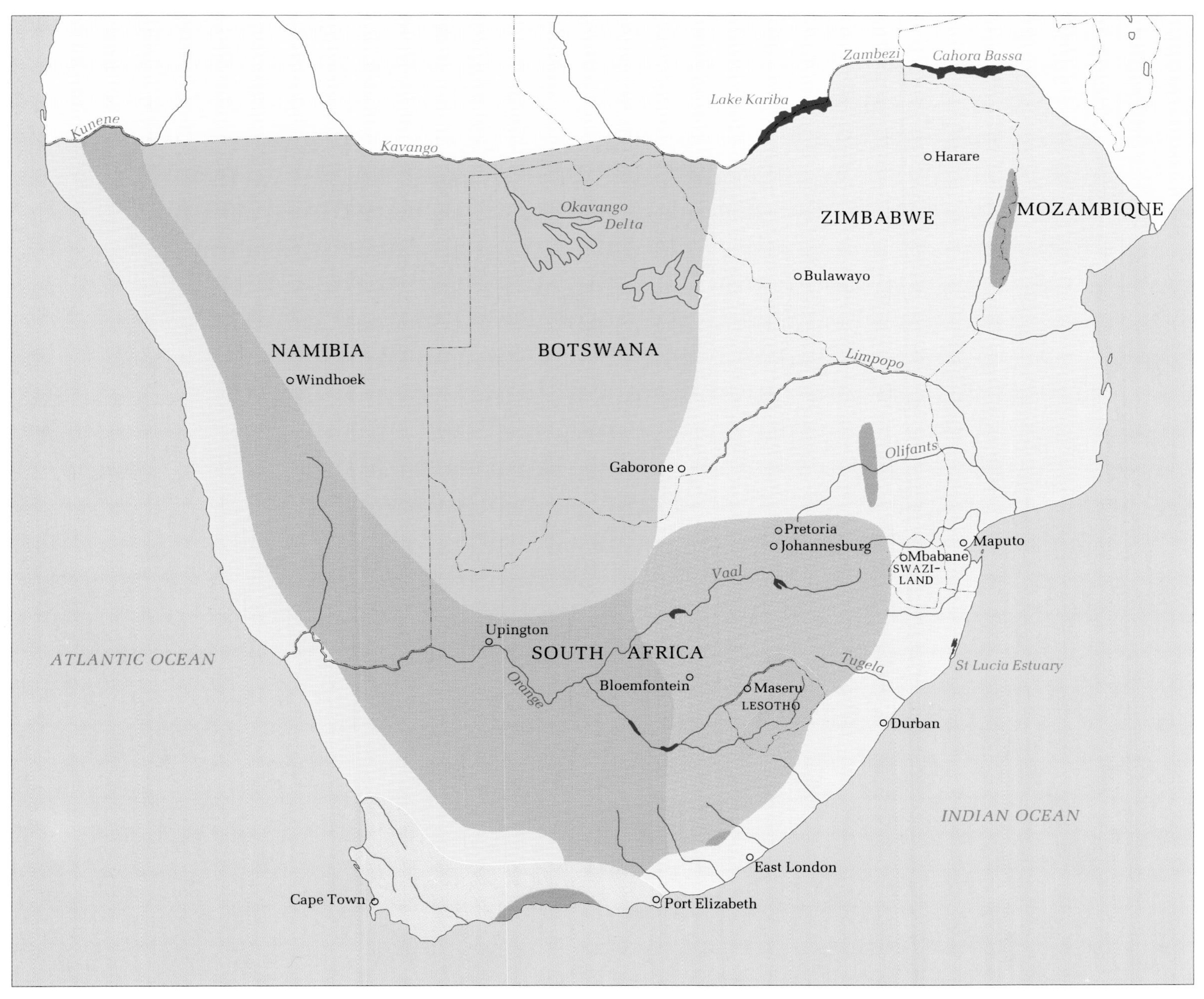

THE NATURAL REGIONS OF SOUTHERN AFRICA

Each of these regions, and the animals found there, represents a chapter in the book – as indicated by the page numbers below.

Land of mopanes where the lion is king

The boldly coloured and beautiful white-fronted bee-eater.

A gentle breeze ruffles the leaves of mopane and marula trees that punctuate the dusty veld as far as the eye can see. Alone at a waterhole, gulping the muddy trickle that means the difference between life and death, an impala ewe hesitates, quivers and raises her head.

She knows that scent. Just a few airborne molecules are enough to trigger an ancient instinct, and send her fleeing to the sanctuary of the open veld in a series of gravity-defying leaps. There she pauses, ears twitching, tawny coat rippling with nervous energy. She's safe for now. Until the lioness comes again ....

This is the southern African savannah, a vast region of yellow grassland and woodland stretching to endless horizons. Encompassing the Transvaal Highveld, parts of northern and southern Natal, southern Mozambique and most of Zimbabwe, its open spaces provide ideal grazing for antelope, and a rich hunting ground for predators such as wild dogs, cheetahs and lions ... implacable links in the food chain on which all life depends.

Classified into two main groups, arid ('sweet') or moist ('sour'), the savannah harbours a huge variety of living things, all of

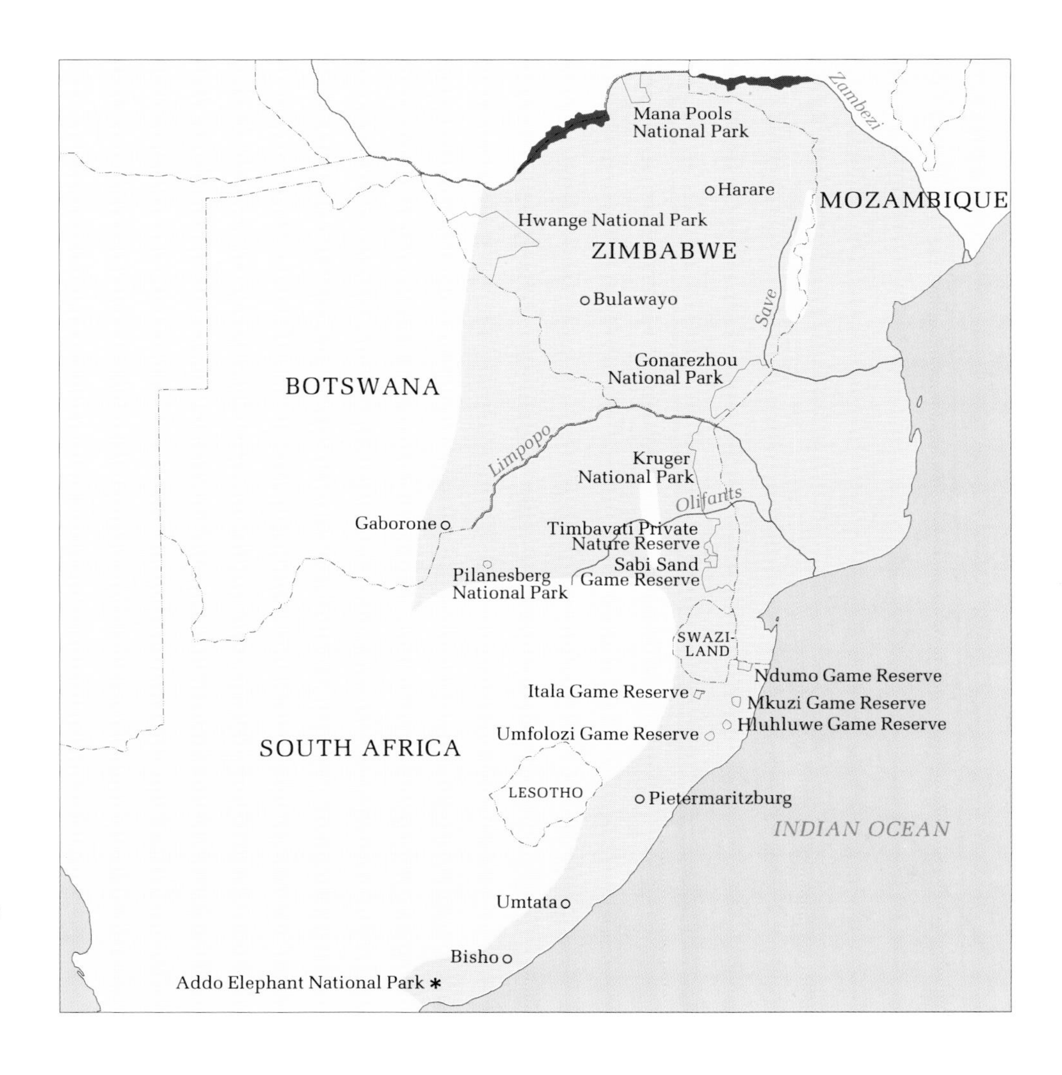

A solitary lioness surveys her territory. Although she prefers to remain with the pride, her hunting expeditions at times require her to wander as far as 45 km from her companions.

which contribute to a delicate balance that we are only now beginning to understand.

Forged in life and death, tempered by æons of wind, rain, drought and sunshine, the ancient relationship between plants and animals continues to work its magic on the savannah as it has done for millions of years. Climatic variations, feeding habits and migratory patterns have changed the very face of this world and every species must adapt or die.

Fire, too, plays an important role in the savannah's never-ending cycle of life and death. One kind of vegetation is swept away by the flames, another takes its place. Seed pods burst in the heat and germinate in the soot-blackened soil to give birth to new plants.

Here the magnificent black rhinoceros, irascible as always, pounds the earth and bulldozes its lunch to the ground with an aplomb that can be displayed only by a creature built like a tank.

Overhead, a martial eagle soars effortlessly on an updraft of warm air, its unblinking yellow eyes scanning the ground for prey. It catches a sudden movement, turns its head, and drops into a steep dive, plummeting out of the sun at 100 km/h in a breathtaking aerial attack.

It's over in seconds. A rush of wind, an anguished squeal and the eagle is airborne again, a dwarf mongoose dangling limply from her razor-sharp talons. The eagle comes to rest on a camel thorn tree, then carefully scans the savannah, before her hooked beak starts working at her meal.

Below her a pride of lions languishes unconcerned in the shade of the weather-beaten tree, as a pair of spotted hyaenas devour the remains of a zebra, snarling viciously and squabbling as they gorge.

In the hollowed-out trunk of a long-dead mopane tree, a bush-baby stirs sleepily in the relative safety of its hideaway. Its huge hazel-brown eyes, dark-circled like those of a chronic insomniac, open briefly and close again. The night is still a long way off and the tree cricket perched on a nearby branch will still be there, chirping its strident love call, when the rising moon signals the start of the nocturnal hunt ....

Inevitably, reassuringly, life goes on.

Carmine bee-eaters outside their nesting tunnels in a vertical river bank.

With its large, liquid eyes, acute hearing and sharp sense of smell, the thick-tailed bushbaby is well equipped to detect the approach of its enemies.

CHILLING HUNTERS OF THE BUSHVELD

Hyaenas are probably among the most disliked and least understood of all the carnivores. Their characteristic grin, chilling 'giggle' and reputation as thieving, cowardly scavengers contribute to an image that is anything but cute or noble. Yet in reality they are fascinating creatures with a complex, even charming social structure.

They often hunt in packs, herding prey and attacking from several directions. However, the brown hyaena is a less efficient predator and relies mainly on scavenging, supplementing its diet with reptiles, birds, small mammals and even insects.

Spotted hyaenas at a giraffe kill. Wildebeest, giraffe and various antelope are among their favourite prey. When hunting in packs, one hyaena will usually go for the quarry's underbelly to slow it, after which it is dragged down.

The male spotted hyaena is usually five to ten kilograms smaller than the female, which may weigh up to 80 kg. She produces two cubs per litter, often in an abandoned ant bear hole.

THE UGLY, GRUBBY SCAVENGER

It has a dissipated air about it, a sort of dignified grubbiness that complements its unalluring features. But the marabou stork is no wimp and commands enough respect to drive off any other feathered scavengers from carrion until it has eaten its fill. In between it snacks on snakes, lizards, fishes, frogs or even baby crocodiles. With a weight of 7 kg and wingspan of 2,5 m, this is the biggest of our storks. It is easily recognised by its large grey or pinkish bill and the white ruff at the base of its bare neck.

Observations by naturalists have revealed a strange and rather rude habit among these birds. Although their legs are actually black, they usually appear white because they're coated with a fine white powder – a crystalline deposit of uric acid resulting from their habit of urinating on their legs during hot weather to cool them.

A marabou stork in silhouette. Despite its large size and awkward appearance, it is a graceful flyer that can soar on thermals for hours.

A hapless butterfly falls prey to a white-fronted bee-eater.

The ubiquitous hoopoe's long, thin bill and conspicuous fan-shaped crest are unmistakable. This bird is equally at home in woodland, savannah and suburban gardens, where it uses its pointed bill to probe for grubs and worms beneath the soil.

A mock charge by an adult black rhino is terrifying, especially since there is no guarantee that it's all for show. But rhinos prefer to avoid conflict, sometimes working off their rage on inanimate objects such as termite mounds.

The white rhino's machine-like mouth envelops a clump of grass, presses it against the stiff lower lip, and slices it through.

A white rhino calf grazes under the protection of its mother. When moving, the calf usually walks in front, guided by its mother's gentle prods. Female rhinos produce a calf every four years or so.

A herd of buffalo from the air. These animals congregate in herds that may number from a dozen to several thousand. They observe a strict order of precedence, and violations are sometimes settled in dramatic clashes of horns.

MENACING POSTURES

Adult male buffaloes are much like small boys when it comes to establishing who's boss. Like their human counterparts, they indulge in a great deal of threatening, pushing and posturing but rarely do each other real damage.

In the buffalo herd a bull tries to intimidate another by holding his head high, nose pointed at the ground, or by turning his body sideways as if to stress his impressive bulk. He may also toss his head or make hooking motions with his formidable horns. If this works, his cowed opponent will indicate submission by holding his head low, with horns back, or by placing his nose under his rival's belly or neck.

But if fighting is inevitable, the two charge each other, heads lowered, to take the impact on the horns. The winner is the buffalo who butts the hardest.

Many hunters have discovered the hard way that a charging buffalo isn't easily stopped. In more than one instance, a wounded buffalo has stalked a hunter and charged at him from an unexpected direction, with fatal consequences.

NATURE'S CHECKS AND BALANCES

As everyone knows, the lion has no natural enemies. So why aren't southern Africa's game sanctuaries crowded with these mighty carnivores? We know that cubs have a high mortality rate and that many adults succumb to disease and combat with other lions, but the real answer lies in Nature's system of checks and balances.

When food and water are plentiful, the lion's prey is spread over a large area, kills become more difficult and the lion population is thinned. But in times of drought, when the herds gather near water points, often in a weakened state, they are easily brought down – and the lions flourish.

This peaceful family scene may change dramatically after a kill, when it becomes a case of every lion for himself. Adults vigorously defend their share of the carcass and many an over-eager cub has been badly mauled in the fray.

Contrary to popular belief, lions are perfectly good climbers and have been known to scale trees to get at carcasses lodged there by leopards. A lion will also escape the heat by climbing a tree and settling on a thick branch, where it falls asleep.

Sprawled in attitudes of utter abandon, a pride dozes away the afternoon. Lions avoid the full heat of the sun and may spend up to 15 hours a day sleeping, jaws gaping wide when the weather is especially hot.

Lion cubs are insatiably curious, with a correspondingly short attention span. Other lionesses 'babysit' them while their mother is out on the hunt.

A lion and his mate. Lions are polygamous, and a male will mate with any female which comes into season. Lionesses are equally free with their favours, displaying a brand of sexual aggression that would be intimidating among humans.

The young lion cub is quite vulnerable in its mother's absence. Aside from the threat of death by disease, accident or starvation, the cub can fall foul of a prowling hyaena or nomadic lion.

The yellow-billed hornbill is a familiar sight in game reserves throughout southern Africa. In the rainy months it lives on a variety of insects, centipedes, scorpions and seeds, switching to a diet of ants and termites during the dry season.

The striking bateleur spends much of its day soaring high in the sky while scanning the ground for prey, which varies from insects to small antelope.

A martial eagle conceals prey from its mate flying overhead. The largest of our true eagles, it is an increasingly rare sight in the skies over southern Africa, and has been listed as a 'vulnerable' species.

A young impala follows its mother nervously through the bush. The single young are able to join their mothers in the herd within two days of birth.

Alerted to a distant predator, an impala herd will merely watch carefully without taking flight. But as the intruder draws closer, they take off in graceful, bounding leaps that carry them up to 12 metres at a bound.

Impalas are browsers and grazers, eating a variety of grasses, leaves and small twigs of shrubs and trees, interspersed with seed pods and wild fruits. They stay close to drinking water but also get moisture from their food.

Impala lambs and their mothers join the herd just 24 hours after birth, the young remaining in 'creche' groups for a while.

Although they do eat grass at times, impala are mostly browsers, feeding on the leaves of bushes and trees, and on wild fruits and legumes. The average male weighs 50 kg; females about 10 kg less.

An impala ram shows a splendid pair of horns. During the mating season, some of the males mark out territories into which they move the ewes. Other males live apart in bachelor herds.

FLEEING BY LEAPS AND BOUNDS

There's little in the southern African bush that can match the grace and symmetry of an impala herd fleeing through the grasslands. When startled, an entire herd will take off, moving over the ground in superbly synchronised fashion, swerving through and bounding high over the bush in perfect unison – a breathtaking spectacle.

These delicate antelope are remarkably nimble – capable of leaping fully three metres into the air over a distance of 12 metres, without apparent effort. In spite of their jumping ability, they do sometimes fall prey to such predators as lions, leopards, cheetahs and wild dogs.

The impala ram's open-mouthed lip-curl is part of the sexual process. This antelope is at his noisiest during the mating season, his loud snorts and grunts more like those of a large carnivore than a bovid.

Herds of impala number anything from 20 to several hundred individuals. All the young are born within a few days of each other – which presents predators with an over-supply of lambs, thus enhancing the survival rate.

A herd of Burchell's zebra from the air. Herds are made up of family groups, the bachelor groups taking up the rear or flanks when on the move.

Among the features that distinguish this species from the mountain zebra are the greyish or yellowish 'shadow' stripes between the black on the hindquarters, and the lack of a dewlap. No two individuals are identically marked.

Shy and nervous, a Burchell's zebra approaches a waterhole with extreme caution, its highly developed senses of sight, hearing and smell attuned to the slightest hint of danger.

COURAGEOUS CUSTODIANS

The magnificent shining coats and rounded bodies of Burchell's zebras galloping across the savannah is one of the truly memorable sights in the community of animals.

Timid, restless and excitable, these zebras utter a piercing whinny kwa-ha! kwa-ha!, identical to that of their close relative, the now-extinct quagga which the Khoisan named for its barking neigh.

The males are courageous. If a predator threatens the herd, the stallions will take a protective rearguard position while the rest of the group flees. In large herds, stallions may form a defensive line along the flanks, and actually attack predators which pose a threat to their young.

Bonding in the bush. Burchell's zebras often associate with blue wildebeest and other antelope, each species benefiting from the increased vigilance and sharing of the threat from predators.

Wild dogs stalk two Burchell's zebra in a thicket. Although timid, these zebras can put up a spirited fight when threatened. In one incident a mare was seen to kill a spotted hyaena with a devastating kick.

HIGH-SPEED HIT SQUADS

A pack of wild dogs on a hunt is an awesome sight. Streaming across the veld behind the pack leader, working like a well-trained hit squad, they run down and consume their prey with shocking ferocity. Implacable and seemingly tireless hunters, wild dogs may continue their chase for several kilometres at speeds of up to 60 km/h, rarely breaking off to take easier prey. Small animals are torn to pieces within seconds, while larger prey are sometimes disembowelled on the run before being pulled down by the pursuing pack.

Wild dogs are particularly solicitous of their young and even the hungriest adult may stand aside to let the pups feed first. After a kill, they will return and regurgitate food for the pups as well as the adults left behind to look after them.

Wild dogs jump and snap excitedly around a tree as a leopard snarls its defiance. Individually they would be no match for the big cat.

COOLING OFF

Because elephants have no sweat glands, water plays an important part in helping them cool down. They drink about 80 litres of water a day, and may suffer physical damage if they go without water for a few days or more. Elephants are capable of fording wide rivers, either by swimming or walking along the bottom and using their trunks as snorkels to take in air.

Elephants wander long distances in search of food, usually guided by the most experienced individual in the herd. Having been that way many times, the leader can locate waterholes and even hidden water sources.

Elephants on the move. Nomads by instinct, their once large ranges have been greatly reduced by human boundaries and encroaching 'civilisation'.

An elephant prepares to eat after breaking down a tree. When restricted to a sparsely wooded area, elephants may cause havoc by stripping and breaking every tree in sight, or denuding them of bark.

Few predators can match the leopard's power and agility. But sometimes even this superb hunter makes mistakes, for example when it tackles a porcupine without due caution. Leopards have been found dead after such encounters.

THE LEOPARD'S CLAW

Like most cats, the leopard is equipped with retractable claws, a most useful characteristic in an animal that relies heavily on its claws to bring down prey. When not needed, the claw-bearing joint is folded back over the preceding joint, protecting the claw with a sheath of skin and preventing it from blunting on contact with the ground. But when the leopard extends its paw to strike, a tendon connected to the muscles in its leg pulls the first joint, moving it forward and down to expose the claw. Driven by powerful muscles, a leopard's paw is a deadly weapon.

Leopards sometimes employ diabolical cunning to secure their food. One hunter recalls watching a leopard rolling in buffalo dung to disguise its own scent, enabling it to creep close without frightening its prey.

Coming this close to a leopard would be impossible without the magic of a telephoto lens. The leopard is a silent and secretive creature by habit, a fact that probably contributes to its survival in areas inhabited by man.

Leopard cubs learn quickly and will often make their first kill at the tender age of five months, usually accompanied by their mothers. They remain with her until they are about two-thirds her size.

The kudu's massive spiral horns give it an air of grace and dignity. These horns, present only in the male, are known to reach 1,8 m in length.

A male and female kudu. The males are fawn-grey while the females often have a cinnamon-coloured tinge, with less distinct facial markings. They also lack the male's beard and fringe down the dewlap.

A juvenile male kudu stands motionless among the trees, ears cocked towards the camera. Although clumsy in flight, the kudu is a powerful jumper and can vault a two-metre fence without difficulty.

MONKEYS PASS THE BUCK

There's something about vervet monkeys that makes comparisons with humans inevitable. For example, although the pecking order within a troop is clearly defined, a challenge to the leadership does not usually lead to a fight between the original protagonists. Instead, when a subordinate monkey is threatened or bitten by a dominant individual, it redirects its own anger to the next one down the line. Similar behaviour has been observed in office environments all over the world ....

A vervet monkey dribbles after gorging on a wild cucumber. These monkeys form strong social bonds and become very distressed if separated from their troop at night. They spend hours grooming each other, usually in the mid-morning.

Protected by the sharp thorns of a bitter aloe, a tree squirrel holds a nut firmly with its forefeet while it extracts the tasty kernel.

Mournful-looking and timid by nature, the African civet rarely ventures out during the day. When disturbed, it often flattens itself to the ground.

A male nyala. These medium-sized antelope often associate with baboons, feeding on leaves and fruit they dislodge from the trees overhead.

The female nyala lacks the impressive horns of the male and is coloured a handsome chestnut, with or without the white chevron between the eyes. Females are rarely aggressive but have been known to butt each other when angry.

Monarch of the glen: a magnificent male nyala surveys his world. Nyala are not territorial and are found in a variety of habitats, though they always seek out areas where thickets provide cover and protection.

Giraffes feed on a thorn tree, stripping off the leaves and tender shoots with their long tongues and supple lips. Usually docile, the world's tallest animal can put up a spirited defence, and can kill a lion with a single kick.

A giraffe chews unconcernedly at a twig, with red-billed oxpeckers in attendance. The oxpeckers perform two valuable services in exchange for their meal: they rid the giraffe of ticks and other parasites, and also warn it of impending danger.

THE UGLY PRIMA DONNA

The tough little warthog, a long-legged member of the pig family, gets its unflattering common name from the wart-like nodules that protrude at the side of its over-large head.

The male has two pairs of these quite unsightly tubercles, the females one, and their purpose remains something of a mystery. Warthogs congregate in small family groups of mother, father, juveniles and up to four piglets. The young warthogs are usually born in summer following a gestation period of about 170 days.

A warthog trots through a forest in search of food. Warthogs subsist largely on grass and roots, which they dig up with their hard snouts, often dropping to their knees in ungainly fashion and shuffling forward to forage.

The curving upper tusks of the adult male warthog are capable of inflicting savage wounds. An adaptation of the canine teeth, these tusks grow to about 60 cm in length. Females, such as this one, are smaller in size and have shorter tusks.

On the move: a warthog trots confidently across the grassy floodplain, its long, thin tail held characteristically erect.

The sensual delight of a morning mudbath. In the hotter midday hours, and at night, the warthog retires to its burrow (which it will enter tail first), a hole often taken over from a departed aardvark or porcupine.

Red-billed oxpeckers feed from parasites that infest the warthog's hide, providing a welcome cleansing service. The birds also function as an early-warning system, their calls of alarm informing their shortsighted host of approaching danger.

A cheetah hurtles through long grass in pursuit of its victim, usually a smallish antelope such as a springbok or an impala. These hot pursuits are often unsuccessful, in spite of the cheetah's speed.

THE HUNT

Animals rely on a variety of remarkable adaptations and strategies in their ceaseless hunt for food in the southern African wildlife community. Lionesses combine stealth, co-operative hunting and ferocious strength to drag down fleeing animals. A pack of wild dogs uses its combined stamina and endurance to run down an exhausted quarry, while a cheetah relies more on speed than endurance to capture prey, often reaching 75 km/h in pursuit of its victim.

A crocodile employs camouflage, floating up to its prey unseen and unheard, before lunging from the water in a burst of terrifying energy and power.

A leopard crouches low as it stalks through grass towards a group of grazing antelope. The spotted hide is an excellent camouflage in all types of terrain.

Young antelope have little or no chance of escaping from a pursuing cheetah. After bowling the youngster over with a front paw, the cheetah will invariably strangle it, sometimes holding on to its throat for several minutes.

An African marsh harrier scavenges a fish from a river in Natal. The harrier eats small animals, mainly rodents and birds, but it will also take frogs, reptiles, birds' eggs and, as the picture shows, fish.

DEATH FROM THE SKY

Resourcefulness and cunning are important factors in the hunting strategies of some animals. In the Kalahari, birds of prey swoop out of the sun on groups of feeding mongooses, so that sentinel mongooses on the ground will not see them; the lightning-quick caracal uses its prodigious reflexes to catch birds in flight; the African marsh harrier tumbles out of the sky with outstretched talons, surprising its quarry on the ground.

Hunting raptors pose a dire threat to some small animals in the wild, such as this suricate, keeping watch on an old tree stump.

A serval pauses on a river bank in its search for food. Great wanderers, these cats eat mainly wild mice and occasionally game birds, but they will also take reptiles and insects. Serval often hunt in watery, swampy terrain.

Tussle for territory on the grassy plains

The intricate black and red patterns on the body of the hawk-moth caterpillar help to discourage the attentions of potential predators.

It's autumn in the rolling Orange Free State grasslands, and the black wildebeest is establishing his territory, marking out an irregular boundary with glandular secretions. And the odorous message to rival bulls is perfectly clear: cross this line at your peril!

Without warning, he wheels and takes flight, snorting furiously, head tossing from side to side – then stops with equal suddenness to peer over his shoulder in apparent bewilderment. Moments later, the rest of the herd bolt in the same direction and repeat the performance. No-one can explain why black wildebeest behave in this strange manner.

Their home is the temperate grassland or grassveld found mainly at higher elevations in the summer rainfall areas of southern Africa, and stretching over the northeastern Cape, most of the Free State and the Transvaal Highveld. These handsome antelope share their grassland habitat with a variety of animals, large and small.

One of the less conspicuous of these, the aardwolf, emerges silently from its borrowed lair to search for termites. Illuminated by the moon, the aardwolf could be mistaken for a jackal, but its

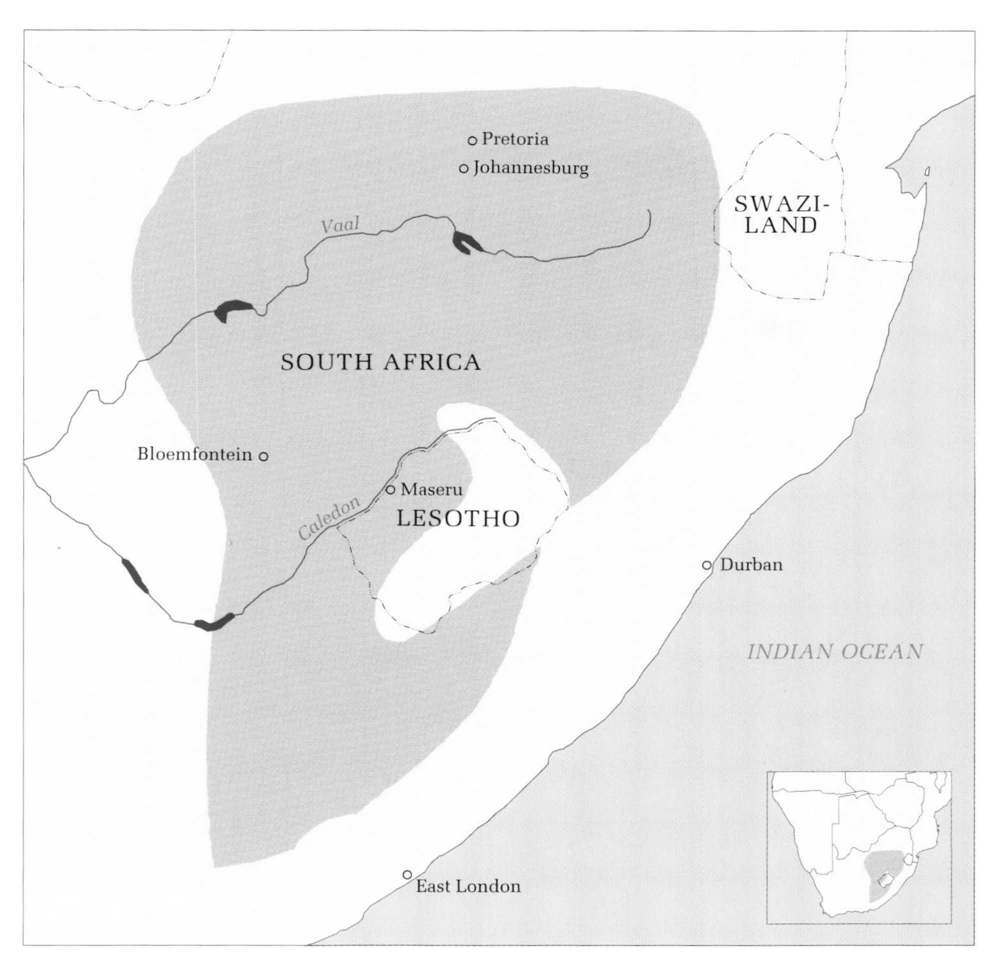

A pair of blesbok pause on the edge of a hill fringed by golden grassland. Blesbok rams mark their home ranges with dung middens. The territorial ram anoints his face in them so that the scent marks him as the dominant male in the area.

hyaena-like gait, and the shaggy tail held out straight behind it, leave no question about its identity.

Moving at a trot along the grassy plains, the aardwolf's ears are pricked and it holds its head close to the ground. Suddenly, perceiving a threat nearby it stops and growls threateningly. The hairs along its back rise to form a mantle that adds stature and menace to its presence.

Deciding that discretion is the better part of valour, the aardwolf does a quick about-turn and darts for the safety of a nearby hole in the ground.

Farther to the north, a herd of blesbok graze on Highveld grasses near a fast-dwindling waterhole. A jealous male stands guard over his harem. Another male, ignoring the warning scents that mark the other's territory, strays closer – and triggers an awesome display of aggression.

In most cases the accent is on bravado rather than physical confrontation, and the exchanges are reminiscent of two schoolboys squaring up for a fight that neither really wants. But this time it's more serious.

Giving its opponent a menacing stare, the blesbok snorts loudly. Unimpressed, the intruder moves in and the two males meet with a clatter of horns. The battle is over quickly as the defeated blesbok dashes away.

The massive leopard (or mountain) tortoise pokes its scaly primeval head out of its black and yellow shell. Weighing up to 27 kg, and growing to a length of 50 cm, this is the largest of southern Africa's land tortoises.

The noise of their clash carries across the veld to the tufted ears of a caracal, perfectly camouflaged in the long grass of its hunting grounds. It's uneasy; this nocturnal hunter is not at its best during the day. But it has two hungry kittens hidden in an abandoned hole nearby and there's an unwary ground squirrel just begging to be caught.

Whiskers twitching, body flattened against the ground, it begins to move in on the unsuspecting squirrel ....

Waiting in the wings – as always – is the black-backed jackal, which trots silently in ever-decreasing circles around the site of this potential kill. For this wily scavenger, no meal is too small.

A sungazer peers skyward from a vantage point. This reptile is covered in horny scales, each of which carries a needle-sharp spine.

A pair of Burchell's zebras rub necks and shoulders in the grassland. These zebras exchange greetings by opening and closing their mouths, without showing their teeth. Vocal communication among these animals is important.

KEEPING PREDATORS AT BAY

Burchell's zebras have a carefully planned strategy for surviving a possibly lethal attack by a pursuing predator: they run at only half their potential speed. This enables the group to bunch together tightly, with the mares and young foals in front, and the stallions at the rear.

Usually this tactic works. A zebra that breaks away or is isolated from the fleeing group is in grave danger of being caught.

Stallions often repel hot-pursuits by hungry predators by giving hefty, defensive kicks with their powerful hindlegs. Such kicks, if connected properly, can easily break a lion's jaw or hurt a predator enough to send it scurrying away.

A group of Burchell's zebras strides through grassland. Foals, which can stand within 15 minutes of birth, eat the fresh dung of adults in the first week of their lives. After this, they start grazing with their parents.

Some wildlife experts believe that a zebra's stripes serve as a mechanism for individual identification. Like fingerprints, each zebra's stripes are uniquely configured.

The more common of southern Africa's two zebra species is named after William Burchell, a talented Victorian artist, naturalist and traveller who helped pioneer research into a beautifully illustrated work on wildlife.

A black-backed jackal peers cautiously through long grass. These irrepressible scavengers are quick to locate sites of a kill. They are commonly found around waterholes, where the large carnivores ambush unsuspecting victims.

THE MASTER SCAVENGER

Jackals are the master scavengers, true opportunists that will resort to almost any tactic to steal a meal. They will snatch a piece of carrion from a vulture's beak or even deprive smaller carnivores of a well-earned meal.

When a pride of lions sets out to hunt, jackals usually follow at a safe distance, trotting behind until a kill is made. Then they sneak in quietly and quickly, the bolder ones hurrying in to snatch the odd discarded morsel.

Their tastes, however, extend well beyond the carrion of large animals: jackals are true omnivores, and will also eat rodents, lizards, insects, birds and their eggs and even the fruits of the veld.

Most jackal pups are born in litters of three, usually in abandoned ant bear holes. They are looked after – fed, groomed, guarded and taught the ways of the wild – by older juveniles as well as by the parents.

A bat-eared fox pup pricks its large, sensitive ears above the ground, listening intently for the noises of subterranean insects. If it hears a movement under the ground, the fox will start digging furiously.

The sprightly little melba finch. The engaging song of this attractive, secretive bird can last up to 16 seconds.

A family group of bat-eared foxes huddles together as the day draws to a close. The cubs are born in the burrows of their parents, and remain there for about three weeks before venturing out.

Eland are the largest of southern Africa's antelope, the males weighing around 700 kg, and the females weighing about 460 kg. These gregarious creatures congregate in herds of up to a thousand.

A black wildebeest warily scans its surroundings for signs of danger. These ungainly looking antelope were once threatened with extinction, but careful conservation measures have restored their numbers.

Proud and powerful, male red hartebeest engage in a furious tussle for dominance of the 'harem'. These antelope normally gather in small groups, but migrations of more than 10 000 have been reported in Botswana.

Moist-eyed beauty in the bush. This bushbuck watches warily for any sign of danger. Its light brown colouring enables it to blend in with its environment. Bushbuck are found in grassland only when this is associated with riverine woodland.

A bushbuck male shows poise and dignity as he surveys his immediate surroundings. Because they are not very good runners, bushbuck try and avoid capture by 'freezing' so that predators cannot see them.

Slim and elegant in its spotted and barred coat, the serval is a rare but beautiful inhabitant of the grasslands. Its preferred habitat is the bushveld, where it lives a mostly solitary life. Servals prefer to hunt at night.

THE SERVAL: SHY LONER

Servals are shy, solitary animals and this, allied with their nocturnal habits, make them hard to spot. They sometimes wander long distances at night in search of prey, usually sticking to well-worn paths or even roads, rather than taking shorter routes across irregular terrain. Rodents are their favourite prey and, unlike other wild cats, servals will penetrate deep into swamps and other wet areas in pursuit of vlei rats.

When threatened, servals flee for tall grass or any underbrush that provides cover, occasionally even taking refuge in trees. One serval was seen clawing its way nine metres up the smooth trunk of a eucalyptus tree to escape a pack of dogs, using its razor-sharp dew claws to grip the bark.

The serval's relatively large ears are accentuated by its small head. Sharp hearing is crucial to the success of its night-time hunts for rodents (rats and mice form the bulk of the serval's diet), birds, reptiles, small mammals and insects.

A cheetah devours its kill. When food is scarce, two or three adults might collaborate to hunt a larger mammal such as a young wildebeest.

HOW ANIMALS GET THEIR FOOD

Nature has endowed animals with highly specialised equipment to help them in their quest for food. The cat family uses its weaponry of tooth and claw to hold down and eat its prey. Lions, for instance, have scissor-like molars (called carnassials), which they use to cut chunks of meat from a carcass, and small incisors, ideal for gnawing at bits of flesh inaccessible to the other teeth.

Cheetahs have much smaller teeth, but use these to good effect when eating their prey. Members of the cat family, including servals and caracals, have tongues covered with hook-like papillae that clean up shreds of meat from the bones of their victims.

As they are rather inoffensive, peace-loving creatures, cheetahs are often chased from their meals by lions or hyaenas. They favour small mammals such as impala, springbok, reedbuck or steenbok.

The beautiful paradise flycatcher takes food to its young in the nest. These flycatchers build exquisite, cup-shaped nests of roots, bark and other material, bound with spider web, which they camouflage with lichen.

A giraffe bull browses on acacia blossoms. The giraffe uses its long tongue to strip shoots and leaves into its mouth.

Servals use their large ears to locate rodent prey, pouncing silently and swiftly in an arc to pin their quarry down.

The lion's weaponry of huge teeth and sharp claws enables it to maintain its position as feared king of the food chain, and, indirectly, to provide scavengers such as black-backed jackals and vultures with food for their own survival.

TAKING THE MEAL

Most large land-predators use their paws or feet to hold their quarry down while tearing chunks of meat off the carcass. Spotted hyaenas and wild dogs, in particular, have powerful jaws to assist them in this task. The teeth of wild dogs are adapted to holding and slicing their prey, which they tear to pieces during the hunt. Wild dogs commonly bolt fresh meat down and regurgitate it later at their dens, at the request of their pups.

A peculiar adaptation of the aardvark is its long, sticky tongue which acts as a natural trap for the ants which make up a large proportion of its diet.

A ground squirrel munches a stalk in the Kalahari. These animals prefer to eat grass, bulbs, roots and seeds, but occasionally include insects in their diet.

Unlike lions, a pack of wild dogs is disciplined and patient when it comes to eating the kill. The adults usually stand back to let the juveniles eat first. Wild dogs will drive hyaenas and even leopards away from a kill.

Secrets of the enchanted forests

A wall-eyed glance from the vigilant eyes of a flap-neck chameleon. It lives on a variety of beetles, grasshoppers, spiders, snails and centipedes.

In the undergrowth, something stirs. A bewhiskered nose emerges from the carpet of damp leaves, twitches furiously and disappears. Seconds later the leaves scatter and a tiny forest shrew darts across the clearing, a grasshopper held firmly in its mouth.

Ever hungry, always on the move, the shrew's runaway metabolism drives it to feats of gluttony that would be impossible in any other creature. A slug goes the way of the grasshopper, then a pair of snails, a mole cricket ....

Such is life in the southern African forest, a world within a world and a place of rare beauty. Forests of varying sizes are found all over the subcontinent but occur mainly along the narrow coastal terrace as long, narrow belts of trees or as dense patches in Natal and the eastern and northern Transvaal.

In the southern Cape, an irregular belt of indigenous forest survives in sheltered valleys, river gorges and mountain kloofs. To the east, the ancient Knysna and Tsitsikamma forests are all that remain of the indigenous forests that once covered huge areas of southern Africa.

A vervet monkey peers from a dense thicket where it has been nibbling on leaves. It sleeps in the upper branches of large trees, where it is hidden from predators by the thick foliage.

The striped polecat has a devastating defence against its enemies, turning its back and ejecting a foul-smelling fluid from the anal glands.

All these forests teem with life, most of it invisible to the untrained eye. A hawk moth rests on the branch of a keurboom, its outspread wings perfectly camouflaged against the coarse bark. It will remain there without moving until sunset, when it will fly off to suck the nectar from moonflowers.

But only if the twig snake doesn't see it first. Unblinking and motionless, its grey-coloured body resembling a dry branch, the snake blends so well into its background that it seems to disappear. Certainly the chameleon, itself no mean hand at disguise, doesn't see it: the snake's head snaps forward and the unwary chameleon is injected with a lethal dose of haemotoxic venom.

Far below, another forest denizen grunts with satisfaction as it wallows in the mud on the banks of a forest stream; the mud layer will protect its hide from insect bites. Toilet completed, the bushpig heads off in search of food, its probing snout leading it to a supply of succulent roots, which it jerks from the soil with a sweep of its sharp tusks.

March is showtime for the black ironwood, with its display of cream-coloured blossoms that turn the forest canopy into a visual extravaganza. Once felled for use as railway sleepers, these trees stand a better chance of survival today because modern sleepers are made from concrete.

But humans are not their only enemy; lichens and other parasites are devastating. Some have deceptively appealing names – traveller's joy, lemon capers, climbing saffron. The strangling twiner literally chokes a sapling to death.

Every tree has a natural role, and many have a place in human history. The hardy stinkwood, often seen 'decorated' with tree club moss, was used to make the tented wagons that carried the Voortrekkers into the hinterland.

In another forest, a Knysna lourie perches on a lemon-perfumed Cape chestnut tree and preens its spectacular red-and-green plumage. The red colouring is another of Nature's marvels: it has been traced to a pigment derived from tiny traces of copper in its diet.

The forest's palette is boundless.

A solitary female bushbuck in the dense bush where it seeks cover during the daylight hours. This buck will readily take to water and swims well. One individual was observed swimming 3 km without any sign of tiring.

Late afternoon and early morning are the preferred feeding times for the common duiker, seen here eating fresh leaves. Duikers also eat flowers, seeds and fruit, occasionally using their sharp hooves to dig for roots and tubers.

THE DUIKER'S HIDDEN STRENGTH

Dynamite sometimes comes in beautiful packages. Take the common duiker, for example: dainty and delicate in appearance, tipping the scales at a modest 20 kg, it is surprisingly strong and puts up a good fight when trapped or threatened. Its sharp horns are used to good effect in close encounters, and even its hooves become potentially lethal weapons.

But fighting is generally a last resort, and the duiker's first inclination is to flee in a series of plunging jumps that carry it across the ground at a surprising speed. Its zigzag flight comes to an abrupt halt in the cover of a dense thicket.

Named after its richly coloured coat, the red duiker is smaller than the common species, and just as shy. It usually seeks the protection of riverine forests, wooded ravines or tree-clad mountain slopes.

DROPPING IN FOR A BITE

The powerful crowned eagle thrives in the forest, where it perches on a high branch while examining its surroundings with sharp eyes. It rarely takes prey in flight, usually preferring to drop on to a bird or animal from its perch, in a flurry of checkerboard-patterned wings.

The breeding pair build their nest of sticks in the fork of a tree in late winter or early spring. Only one of the two chicks survives after hatching, remaining with its parents for up to a year before achieving full independence. This species of eagle is very vocal, the male calling loudly as it soars over its territory.

A pair of wood owl chicks display the fluffy plumage that characterises all young owls. Usually occurring in forests, especially dense riverine forests and exotic plantations, they roost close to the trunks of large trees during the day.

Winged monarch of the forest. The crowned eagle's relatively short wings are well suited to its forest habitat, where it can climb almost vertically with prey such as vervet monkey or blue duiker clutched in its powerful claws.

Vervet monkeys are active from dawn and will forage until mid-morning, when they take a break and seek out a shady spot to escape the baking heat. They return to their sleeping places in the trees before sunset.

The tree is home, boudoir, refuge and larder to the vervet monkey. It eats fruit, seeds, flowers, leaves and insects with equal enthusiasm but rarely swallows its food immediately, usually waiting until its cheek pouches are full.

A family group of vervet monkeys. They use many different calls to indicate hunger, alarm, pain, distress or aggression. The sound varies from a gargle to a harsh stutter, usually accompanied by a distinct facial expression.

MOTHERLY LOVE OF THE MONKEYS

Young vervet monkeys are treasured by the females in the troop and their fiercely protective maternal instinct is one of their most endearing features. Should a strange juvenile be introduced into a troop, the females will immediately surround it and pick it up in their arms.

So strong is this instinct that even a three-month-old female may pick up and caress a baby monkey. While the males are generally indifferent to infants, their presence seems to strengthen the bonds within the troop, which acts as one in protecting them from predators.

Tiny and vulnerable, sucking its thumb in a disconcertingly human fashion, a baby vervet monkey surveys its world. It attracts its mother's attention with a gargling noise or a drawn-out, staccato cry.

Chacma baboons often sleep huddled together, guarded by one or more adult males who bark and squeal at the approach of a predator. Most troops spend the night in the relative security of rocky hills but others opt for tall trees.

This bushpig's warning glance says it all: don't come too close. These cautious and cunning omnivores are dangerous when cornered or accompanied by young, and their impressive lower canines can cause serious injury.

RUMBLES OF CONTENT

Elephants are active during the day as well as the night but will usually spend the hottest part of the day resting or sleeping in the shade of a large tree, occasionally fanning themselves with their big ears.

Elephants communicate by several means, one of the more remarkable being the deep rumbling noise once thought to originate in their stomachs (in fact the sound is produced by their vocal chords).

Reminiscent of the noise usually associated with indigestion, it is best heard during the stillness of the night in close proximity to a feeding elephant.

Elephants move at a slow walk between feeding sites but can achieve quite a speed when charging an enemy or rival. They continually test the air with their trunks, alert to the presence of predators which may threaten their young.

Dark marks on the elephant's head indicate the location of temporal glands, which secrete a pungent-smelling fluid with no apparent purpose. The secretions occur in both sexes at any time, irrespective of age.

Green algae forms the perfect camouflage for a large crocodile at Lake St Lucia. Such camouflage effectively enables the crocodile to drift unseen towards animals drinking at the water's edge.

The crested francolin's streaked and dappled brown colours blend naturally with the colours of its environment. The francolin nests on the ground in a scrape lined with grass and surrounded by leaves.

An eye betrays the branch-like form of a Scops owl in a tree. When it is disturbed, the owl usually elongates the body and closes the eyes almost completely, so that it is more or less indistinguishable from a tree trunk.

ART OF THE UNSEEN ANIMALS

For many animals camouflage – the art of merging unseen into the surrounding environment – is the master adaptation in the great game of survival. It enables large animals to creep up and catch prey undetected; and, just as important, it enables smaller animals to escape detection.

But to many species, camouflage is not something that has come easily: in some instances it has been a physical adaptation that has taken millions of years to perfect; and mother nature has worked hard, combining sunlight, colours, shades and shadows to provide an ideal camouflage.

The camouflage of this praying mantis is so good that it appears as an appendage to the stalk on which it is standing. Mantids alter their familiar form by ruffling out their wings to resemble leaves.

The spots of a leopard disrupt its shape and form so well that its intended victim cannot recognise it. Leopards often drag their prey into trees, where they can devour them at leisure, without the interference of other predators or scavengers.

Hidden by the trunk of a dead tree, a leopard on the ground merges with the colours of the grassland around it, camouflaging it from potential prey. The base colour of a leopard's coat varies from a golden colour to pale yellow.

Aerial acrobatics among sentinels of rock

Lunchtime in the mountains. A black eagle scans its rugged domain for dassies and other prey.

From the black eagle's leaf-lined nest, the view seems to go on forever. To the east, a line of stunted pine trees marks the edge of the lowlands. To the west, the jagged peaks of the Natal Drakensberg surge through a shroud of sombre grey storm clouds.

Deep in the large stick nest, a single chick eager to fly, stretches its magnificent wings as it gulps down the remains of a francolin killed by its mother. High above, an alpine swift circles and swoops on unsuspecting insects in a stunning display of aerial acrobatics. Suddenly it zooms into a cleft in the cliff face, and finds its nest – a bowl of downy feathers, glued together by its own saliva.

Such is life in one of the largest mountain habitats of southern Africa ... where cobalt-blue peaks pierce the skyline, towering above a world of wildlife especially adapted to freezing temperatures, rocky terrain and precipitous slopes.

This rocky environment includes the massive, fortress-like cliffs, high plateaus and deep valleys of Lesotho, and the Natal and Transvaal Drakensberg. In the southwest, the peaks of the Cape Fold Mountains run from the Cederberg via the Hex River Mountains as far as the Hottentots Holland, and then further eastwards for 600 km almost to Port Elizabeth.

High up in the Cape Fold Mountains, a klipspringer mounts a projecting rock and freezes as if he's posing for a photograph. He's superbly equipped for his habitat, with built-in shock absorbers on his hooves and a coat of special hair to insulate the body from the fierce summer heat radiating off the rock face.

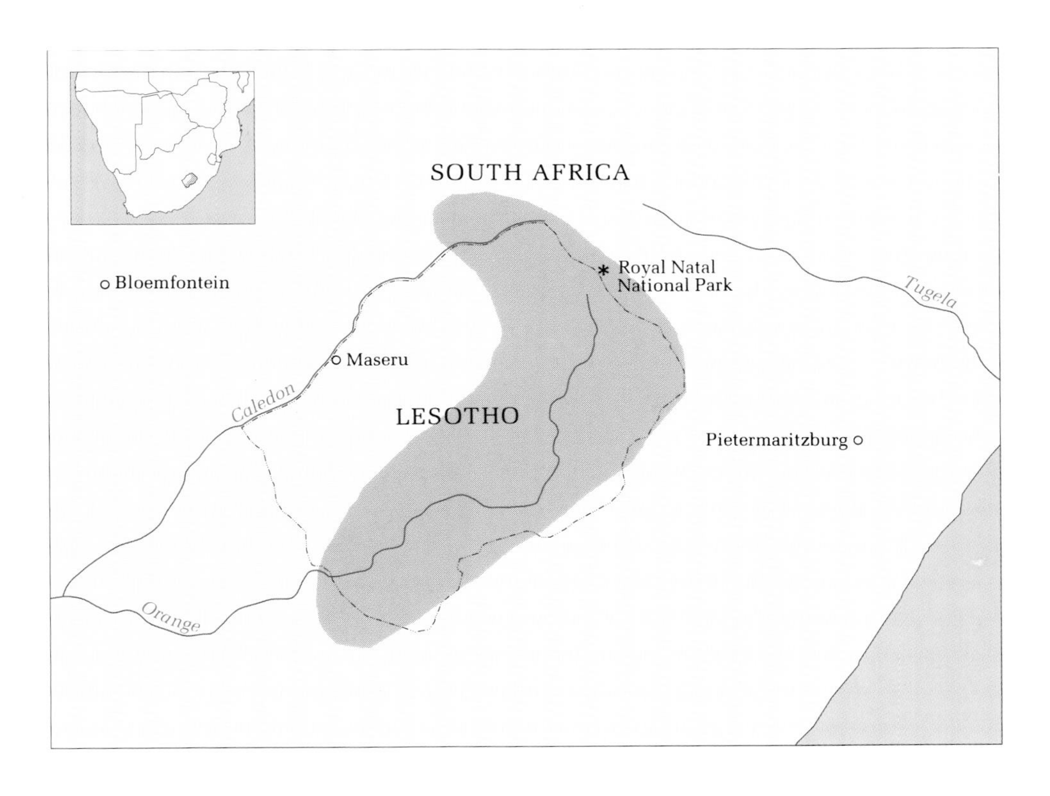

A black eagle soars over the rocky cliffs and gorges that provide its food and shelter. In flight this large raptor is easily recognised by its jet-black plumage, enhanced by white patches on its wings and a distinctive V marking on the upper back.

A rock pigeon ruffles its feathers as it settles down on its nest of sticks.

Even his colouring – from yellow to grey, speckled with brown – is a useful camouflage in the rocky terrain where he searches for the shrubs that keep him alive. Territorial by instinct, he marks his kingdom by rubbing preorbital scent glands on twigs.

A colony of dassies basks in the morning sun, a female sentinel keeping watch for predatory eagles. To counter this threat from the air, Nature has given the humble rock dassie an unusual defence: a thin, movable eye membrane that shields the pupils and allows the animal to look directly into the sun. Unlikely though it sounds, this animal's nearest relative could be the elephant – its front teeth resemble vestigial tusks and, like the elephant, it has blunt plates for nails.

The rocky slopes of the mountain environment are ideal for the chacma baboon, which seeks out such delicacies as grubs, scorpions, termites and snails, and finds shelter in caves and beneath overhanging rocks.

Baboons are continually on the move, ever-alert to the threatening presence of leopards, their arch-enemy. No match for the powerful predator as individuals, a large troop of baboons, acting together, is quite capable of tearing a marauding leopard to pieces.

Most animals will draw on their most lethal resources to defend themselves. While the baboons of the Drakensberg wage their own war against leopards, a battle of a different kind is taking place in a small arena in the eastern highlands of Zimbabwe. A hairy thick-tailed scorpion, largest and most venomous of southern Africa's scorpions, has been flushed from its lair beneath a rock by a hungry gecko. In spite of its formidable weaponry, the scorpion is no match for its opponent, and after a brief scuffle and a few spasmodic twitches, it ends up being devoured by the stealthy gecko.

The southern African hedgehog spends most of the day sleeping in a hole or under cover of a bush, and emerges after sunset to search for food.

Rock hyraxes (dassies) sunning themselves. Although somnolent by nature, they are surprisingly agile animals and their thickly padded feet, moistened by a glandular secretion, can carry them up and down steep, smooth rocks with ease.

HIGH-JUMPER OF THE MOUNTAINS

The klipspringer is a remarkable athlete, able to jump from rock to rock and bound up steep outcrops in apparent defiance of gravity. It is also very cautious, as befits an animal with such powerful enemies as leopards and hyaenas.

Both sexes mark out their territory with scent, the male usually 'overmarking' the traces left by the female. The marking is done via a gland in front of the eye (preorbital gland).

A female klipspringer rests in the shade during a hot summer afternoon. If she spots a predator approaching, she freezes and continues to watch, issuing a high-pitched warning call if it comes closer.

A typical klipspringer pose: motionless, with every sense alert to threats from above and below. This graceful antelope will sometimes browse on level ground but always returns to the sanctuary of its rocky habitat at the first sign of danger.

SOCIAL BONDS AMONG THE BABOONS

Dominant male chacma baboons weigh up to three times as much as the average 17 kg female. These males head the baboon hierarchy. They are responsible for enforcing discipline, keeping watch for predators, and preventing the young from straying too far from the troop.

Social bonds are reinforced by constant mutual grooming, during which fleas and other parasites are located and discarded. A female baboon will carry and carefully groom her infant for three months, not allowing other females to pick it up until it can walk. Young baboons are suckled for about six months.

A female chacma enjoys the attentions of a friend, while her youngster sits within her arms. The bonds of motherhood are so strong that bereaved females have been seen clutching the bodies of their infants days after their demise.

A chacma baboon takes a siesta. The baboon's yawn is not necessarily a signal that it is tired. The yawn, accompanied by bristling hairs, usually means the chacma is threatening something or someone – a possible predator or a human being.

A young leopard climbing a tree. As an adult it will perform the same feat with a full-grown antelope in its jaws.

A blue-eyed leopard cub surveys its world. Cubs are born in caves, holes in the ground, hollow trees and other sheltered places. They are weaned at about six months, after which their mother carries food to them.

A pensive leopard resting on a tree branch, amber eyes not missing a thing. Its whiskers and the two or three extra-long hairs in each eyebrow help it to avoid obstructions during frequent nocturnal wanderings.

While no two zebras have identical markings, the habits of Cape mountain zebras are very similar. On cold winter mornings they move one by one to the east-facing slopes of their territory to absorb the warming rays of the sun.

BACK FROM THE BRINK OF EXTINCTION

Once threatened with extinction, the Cape mountain zebra now flourishes in selected South African reserves. These animals are found in herds of up to 12, but usually number around five or six, consisting of a dominant stallion with mares and foals.

Young males, preferring the freedom of bachelor life, team up together at the age of two years and leave the herd, returning when they are old enough to take on a herd, and harem, of their own. The quest for leadership is often a vicious and pugnacious one in which the challenger or challenged is severely injured.

When danger threatens, the male will utter a high-pitched screech of alarm and hold a defensive position as mares with young beat a hasty retreat.

A Cape mountain zebra in its natural surroundings. Single foals, born mostly in December, nibble at pellets of their mothers' faeces. These contain intestinal micro-organisms which help them digest grasses that form their staple diet.

The elephant shrew derives its name from its trunk-like snout. It is an insatiably hungry and incurably nervous creature.

A Cape mountain zebra foal, its stocky body looking curiously out of proportion. Very young foals are tolerated by other mares within their group, and an orphan is occasionally adopted by another mare with offspring of her own.

The rare bearded vulture has developed a clever way of breaking up large bones into pieces small enough to swallow: it picks them up and drops them on to flat rocks, sometimes repeating the process until the fragments are the right size.

Elephants are the champion drinkers of the world. At the waterhole they take vast quantities of water into their trunks and then squirt it into their mouths, perhaps saving some to squirt over their bodies.

THE SEARCH FOR WATER

From an elephant bull, which can consume a bathtubful of water on a single occasion, to a gemsbok, which can survive indefinitely without lapping up one drop, the liquid needs of animals vary considerably.

Elephants need an ongoing supply of water to survive. When it's hot, adults may visit a waterhole several times a day, but in cooler months they may stay without water for up to three days.

Gemsbok, on the other hand, survive without water, thanks to their adaptive physiological processes. Not only have they a well-developed renal capacity to reabsorb water in the kidneys, thus excreting very concentrated urine; but they can also allow body temperature to rise during hot days and dissipate this heat at night when ambient temperature falls. This process enables gemsbok to avoid sweating and the consequent unnecessary loss of valuable body fluid.

Two Burchell's zebra fill up with a cooling drink at a waterhole. Very rarely far from water, Burchell's zebra, like most mammals, transfer water to their mouths by suction. Typically they like to wade into the water for a drink.

Apart from their need to drink, rhinos visit waterholes regularly to mudbathe, a cosmetic indulgence which helps rid them of unwanted parasites. Like other animals, rhinos are very wary when approaching water.

In areas where water is scarce, such as the Kalahari, lions can go without water for long periods. Where water is abundant, however, lions will drink regularly, taking many minutes to perform the task.

SURVIVING ON DROPLETS

When water is scarce, some animals employ remarkable strategies to secure it. In the northern Cape and Kalahari, the male Namaqua sandgrouse gets liquid nourishment to its chicks by wading into a waterhole, wetting its breast feathers, and flying back to the nest – sometimes a distance of up to 30 km – where the chicks strip the water from the feathers.

In the Namib Desert, the sidewinding adder exposes its body to banks of fog rolling in over the dunes, then slides its mouth across its scales to suck the accumulated droplets. In the same environment, the black-backed jackal licks rock surfaces dampened by fog.

Cats such as lions, leopards and cheetahs, lap up their water, using their tongues as spoons to flick the liquid to the back of the mouth before they swallow it every fourth or fifth lap.

Pigeons have forsaken the traditional 'dip-and-lift' way that most birds drink. Instead, they plunge their beaks into water and suck it in.

Rock pigeons, found over most parts of South Africa, feed their young 'pigeon's milk', a nutritious secretion from the lining of the crop.

A Namaqua sandgrouse's feathers are adapted to retain moisture which it can take to its offspring.

One of the lioness's important functions is to lead her cubs to water, where they take a momentary break from bouts of suckling.

Lionesses savour the cool sustenance of fresh water. They are not very efficient drinkers and may take a long time to quench their thirst, flicking water to the back of their mouths before swallowing.

Dance of the springbok on the plains of the past

Long ears pricked for sounds of danger, a Cape hare makes a rare daytime appearance to search for food.

Luminous clouds of dust curl like a golden fog from the frigid plains of the Karoo, as a column of springbok moves slowly northwards. Suddenly, the symmetry of this moving tide is halted as two handsome males break ranks in the shadow of a flat-topped koppie.

Then, backdropped by the warm afterglow of a Karoo sunset, the springbok begin a strange dance. Heads and haunches lowered, backs arched, they extend their stilt-like legs and bring their hooves together, taking off in a series of extraordinary balletic leaps.

For a brief, magical moment, they appear to be hanging weightless in mid-air, then they return to earth stiff-legged, and bound up again without a second's pause. It's called stotting, and it catapults a springbok three metres into the air.

Hot days and chilly nights, stunted shrubs and desiccated trees, koppies scattered with sun-blackened dolerite rocks that jut from the dusty veld like the abandoned playthings of a long-dead giant ... all of this is the strange and wonderful geographical region called the Karoo.

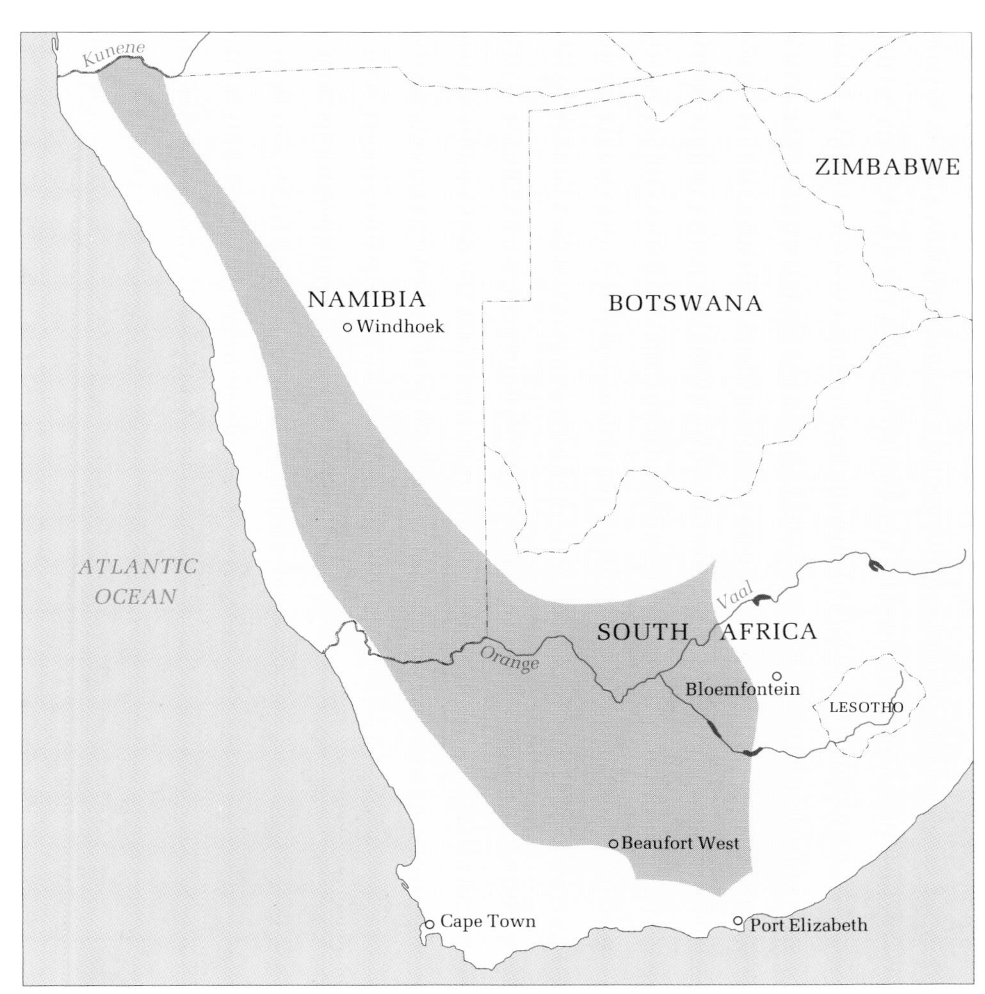

Springbok on the run. When alarmed, these graceful animals can reach almost 90 km/h on the plains. But in spite of their speed, they are often no match for fleet-footed predators.

This fascinating stretch of thirstland, with its ancient layers of shale, silt and sandstone, is actually divided into two distinct chunks: the Little Karoo is located between the parallel mountain chains of the southern Cape, the Witteberg-Swartberg and the Langeberg-Outeniqua ranges. The much larger plateau of the Great Karoo lies to the north of the mountains, extending from Laingsburg to Somerset East.

The Karoo is enormous, and it's getting bigger. Having already swallowed more than half the Cape, its tentacles are reaching into Namaqualand, Bushmanland, the Namib, the Kalahari and even the Orange Free State Highveld.

Among the less conspicuous of the Karoo's fascinating population of living creatures is the dung beetle, which spends a good part of its life gathering dung, patting it into a round ball, and laboriously rolling it along the ground with a furious kicking movement of its back legs.

It then buries the ball in an underground burrow, entices a female to join it there, and then mates with her. She in turn produces an egg right next to the dung ball, and the emergent grub finds itself with a natural and nutritious supply of food.

Fearsome in appearance, with a snake-like tongue, the rock leguaan is itself vulnerable to a ground or aerial attack, though some winged enemies are known to have come off worst. It sometimes feigns death in a bid to escape predators.

The large-spotted genet often climbs a tree when hunting, to improve its chances of locating prey. On capturing a rabbit or other small mammal, it may fall on its side and rake the victim with its back claws, much like a domestic cat.

Completely oblivious of the dung beetle's reproduction cycle, a Karoo padloper plods across the bone-dry soil in search of a succulent coral aloe. Today the tortoise's reputation as a harbinger of rain is justified: the skies open and a sudden thunderstorm envelops the veld, the raindrops creating little craters in the dust until they finally meet in a torrent that erodes yet another layer of precious soil.

The rain brings sustenance and revival as creatures large and small continue their battle for survival with renewed dedication. Orb-web spiders, centipedes and toktokkies go about their mysterious business as they have for millennia; black-backed jackals scurry across the veld in search of carrion or small game; and up on the koppies and in isolated clumps of acacias, black-shouldered kites, steppe buzzards and kestrels wait silently to mount their aerial attacks.

History litters the Karoo with tantalising clues: the fossilised bones of a long-extinct freshwater fish, dinosaur ribs embedded in rapidly decaying sedimentary rock, the trunk of a petrified tree on a rock-strewn koppie.

Sprawled on its bed of twigs in an attitude of utter contentment, a Cape fox remains alert to the slightest sign of danger. When angered it spits and growls, the elevation of the raised tail acting as a barometer of its excitement.

A pair of juvenile Cape foxes go exploring. Little is known of this species' reproductive habits but it is believed that the young – their litters average two to five – are born in the early summer months.

The Cape fox will eat just about anything. An analysis of its diet reveals that it consumes small antelope, rodents, lizards, snakes, birds, eggs and scorpions with equal enthusiasm. It even snacks on wild fruit and green grass.

... AND NOT A DROP TO DRINK

Springbok are usually found in the dryer regions of southern Africa, where they can survive without water indefinitely if their food has a high enough moisture content. However, they will drink readily when water is available and can use even highly mineralised water that other animals would find unpalatable.

Occasionally they eat fruit or dig up succulent roots from the dry Karoo soil to supplement their moisture intake. The springbok is a hardy species, able to adapt and survive in harsh conditions. Its choice of habitat is determined by a number of factors, including the density of the area's vegetation and the availability and mineral-content of the grass, shoots and leaves.

Territorial male springbok do their best to contain females within their home ranges by herding any strays. Defence of these territories is usually more show than fight, although it may involve sparring with the horns.

The springbok's facial markings resemble hastily applied eyeshadow. The male's horns are larger than the female's.

A startled springbok soars gracefully over the Karoo scrub in a series of leaps that seem to defy gravity. The sight of a springbok in flight is one of the truly awesome spectacles of the southern African wild.

A male and female yellow mongoose in typical reconnaissance posture in the arid Karoo where they flourish. These captivating creatures can be quite gregarious, occasionally sharing a warren with up to 20 individuals.

The nocturnal aardwolf loves termites. It locates them by scent, then laps them up with its wide, sticky tongue. This creature resembles the hyaena, but its facial muscles and jaw are not nearly as powerful.

A white or square-lipped rhino with calf. A pair of courting rhinos create an interesting spectacle as they jostle and spar, the male trying to keep the female within his territory while she retaliates with snorts and grunts.

HOW RHINOS SAVE THEIR HIDES

With animals as big and powerful as white rhinos, it makes sense to avoid conflict wherever possible. So when a territorial bull wanders into another's home range, it generally avoids serious confrontation by leaving as quickly as possible.

There may be a token charge or two and a little dust-raising, but as a rule the matter is settled without violence. But the rules change in the presence of a rhino cow in season, when serious fighting may erupt between trespasser and territorial male. Either or both may be wounded by jabbing horns or receive internal injuries from the impact of shoulder battering. Sometimes the clash ends in death.

The white rhino's horns consist of a tight mass of filaments similar to hair, and are not part of the skull. The curved lower horn, which may grow to an astonishing 158 cm, is usually much longer than the one above it.

A suricate family group sun themselves. Ever-vigilant to any signs of danger, a sudden threat prompts a shrill warning bark which sends the whole colony scurrying for the sanctuary of the burrow.

THE VORACIOUS COLONISTS

Suricate colonies may consist of 30 or more individuals who live in warrens with many entrances. They are aggressive squatters by nature, often driving out inhabitants and occupying warrens originally excavated by ground squirrels, who sometimes share them with the yellow mongoose.

Suricates are busy and voracious eaters, often stopping to turn over stones, scratch through debris and dig into the soil for the insects on which they thrive. Their diet includes worms, insect larvae, lizards and even small venomous snakes, which they kill by crushing the skull with their jaws before eating them.

Suricates make good parents: the male diligently guards his offspring while they are suckled by their mother. She may produce more than one litter within a year, though this is very rare. Her young are born in a burrow.

A Burchell's zebra with her foal. These free-ranging zebras depend on readily available water but are catholic in their culinary requirements.

Burchell's zebra at a waterhole. While they will happily paddle during drinking sessions, they are not fond of swimming and will avoid taking to the water even when threatened. They are far happier rolling in dust.

Green grasses constitute the major part of a Burchell's zebra's diet, but leaves, herbs, seed pods and wild fruits may also be consumed. These zebras are fiercely protective of their young, and may risk their lives to defend them.

THE CAUTIOUS SCAVENGER

The black-backed jackal quite deserves its reputation for caution and cunning, thanks to its sharp vision, acute sense of smell and well-developed instinct for self-preservation. In Botswana, one jackal was seen following the scent of a dead springbok more than a kilometre from where the carcass lay, and it is possible that it had been tracking the odour from even further away.

Attempts to trap these wily animals are often unsuccessful because they are highly suspicious of artificial constructions. They will circle the bait and examine the trap carefully before moving off at their characteristic trot.

A black-backed jackal with a captive ground squirrel. It's not a picky eater and will take any prey, from small antelope to rodents, hares, reptiles and insects. Large antelope are consumed in the form of carrion.

A black-backed jackal drinks from a river. A noisy animal at the best of times, this jackal becomes even more vocal when the females are in season. This behaviour is thought to be linked to the establishment of territories.

Tiny miracles of life in endless seas of sand

A black-backed jackal pup explores its desert environment. When very young, it eats food regurgitated by its parents.

Curling its legs inwards, the spider tumbles down the face of a steep sand dune, rolling to a stop at the bottom. Then, as if cartwheeling down a hill were the most ordinary event in the world, it nonchalantly uncurls its limbs and hurtles off – almost running into a small desert lizard.

Rearing up and waving its legs, the spider presents such a fearsome sight that the lizard scuttles off, unaware that it has just witnessed a performance of the dancing white lady – arguably one of the desert's most accomplished entertainers.

Dancing spiders, ants that survive on the nectar secreted by other insects, prehistoric plants with taproots that suck moisture from the desert floor – these are just a few of the extraordinary life forms that inhabit the scorching sands of the Namib, southern Africa's only true desert.

This starkly desolate yet beautiful region, between 50 km and 100 km wide and stretching down the west coast of Namibia and into South Africa, sustains a large variety of living things. From the hardy Welwitschia mirabilis, *an ancient plant which can survive for up to a thousand years, to multicoloured seas of*

A sidewinding adder leaves its distinctive trail across the rolling sands of the Namib Desert. It ambushes lizards and other prey by burying its body in the sand and allowing the tip of its tail to protrude as a decoy.

A common barking gecko emerges from its steeply angled burrow.

lichen clinging to sunburnt rocks, the life forms of the Namib are Nature's own miracles of survival.

*On the dune faces, Grant's golden mole, 8,5 cm long and totally blind, scurries along, pouncing on crickets, beetles or geckos. Nearby, a Tenebrionid beetle (*Cauricara phalangium*), believed to have amongst the longest legs of any beetle in the world, unfolds its stilt-like limbs to bring it as high as possible off the ground. This elevation not only protects the beetle from the scorching ground temperatures, but also exposes it to the relative coolness of the desert breezes, and enables it to gather moisture from fog moving inland.*

Further across the unforgiving sands, a 20 000-strong colony of desert ants relentlessly searches for the source of its favourite diet – scale insects which secrete the sugar-rich honeydew on which the ants survive.

In their ongoing struggle for survival, many animals have developed amazing adaptations to combat the shortage of water: the dune beetle Lepidochora discoidalis, *with the precision of a civil engineer, excavates a metre-long trench to trap water.*

Along the dry riverbeds near the coast, desert elephants dig deep into the dry sand with their trunks until they find water,

Yawning broadly, a Cape fox pup greets the dawn. Its mother alerts it to danger with a sharp bark, spitting and growling at the approach of a predator. She bears anything from one to four pups after a gestation period of about 51 days.

while gemsbok also dig holes with their hooves for a trace of moisture that will bring cool relief.

Plants in this arid region have adapted to survive and propagate: some counter the rapid transpiration by growing smaller leaves; others, such as Welwitschia mirabilis*, rely on their huge leaves to absorb the morning dew.*

The narra plant, with its melon-like fruit, discards its leaves and grows thorns in their place. Arthraerua leubnitziae*, another common Namib Desert plant, survives long periods of drought by shrinking, whereas the Ammocharis lily hoards its water supply in large underground bulbs.*

The Namaqua chameleon, a desert beauty distinguished from other chameleons by the crest of knob-like tubercles along its spine. It feeds on a variety of insects, capturing them with its immensely long tongue, tipped with a sticky pad.

BRASHNESS AND BEAUTY

Springbok are among the best groomed of all our mammals, their multi-tinted coats invariably immaculate, every white and cinnamon-brown hair in place.

When not browsing on shoots and leaves or grazing on fresh grass, they can usually be found attending to their toilet. This can be a complicated affair. They nibble and comb their hair on the sides of their bodies with their incisors, rub with the horns, and scratch themselves with the hind hooves.

However, the result is very impressive and probably accounts for the brash confidence of the courting male, which approaches the female with cocked ears and a bouncy trot that says quite clearly: Look at me – aren't I beautiful?

Springbok seek shelter from the blazing summer sun in the shade of a tree. Their tails are always moving, flicking rapidly from side to side like miniature propellers, and increasing in tempo when danger threatens.

In the dry season springbok obtain valuable protein, phosphate and other minerals from trees and shrubs that appear desiccated by the relentless sun. Sometimes they supplement their diet at mineral licks.

COURTING TROUBLE

A disagreement between a pair of adult black rhinos can be an awesome spectacle. Picture the two animals squaring up, each weighing 1 000 kg and built along the lines of a tank ....

As happens so often in Nature, the really serious fighting usually occurs when two males are after the same female. At first it may be a battle of nerves in which they charge at each other with screams, or try to out-stare each other. But if this doesn't work, the fight could become deadly.

A black rhino in profile, displaying the prehensile hooked lip that accounts for its alternative name. This species is smaller and much darker than its square-lipped cousin and carries its head higher.

MASSIVE ENGINES OF DESTRUCTION

The habitat of the African elephant is a delicate thing, and vulnerable to several influences. Restricted movement, the encroachment of human settlements, long periods of drought, over-population – any one of these could tip the balance with disastrous results.

In fact, elephants are second only to humans in their capacity for destroying their habitat. They break off branches, pull up saplings by the roots, push over bigger trees and strip off bark to satisfy their voracious appetites.

Elephants can adapt to the harshest conditions. These 'desert elephants', though physically the same as other southern African elephants, survive along watercourses passing through the Namib. They travel far in search of food and water.

Gemsbok show grace and dignity in flight across the sandy wastes of the Namib. Both males and females use their rapier-like horns a great deal to spar among members of their own sex.

A group of gemsbok in the Namib at sunset. They dig for succulent bulbs, rhizomes and roots, and in times of drought can somehow locate the buried remains of shrubs even when these plants have left no visible trace above ground.

Porcupine quills embedded in its face, chest and paws, this lion has learned a painful lesson. Hungry predators sometimes circumvent the porcupine's formidable defences by knocking it over to expose the vulnerable underparts.

A sleeping lioness. She may wander deep into the desert, following watercourses frequented by springbok, gemsbok and other prey.

Lion cubs remain with their mother for about two years, sometimes longer. Their hunting skills are acquired by observation and practice, and they quickly learn to respect the lethal weaponry of larger prey such as buffalo and giraffe.

Adult giraffes have no serious predators in Nature other than lions. Attacked, they will defend themselves vigorously, stamping their forelegs like clubs.

The male giraffe has better developed horns than the female and sometimes grows a third horn on the forehead.

A pair of 'desert' giraffe at the northeastern edge of the Namib dunes. Although not a recognised subspecies, these giraffe are special in that they have adapted to the region's arid conditions. They obtain most of their moisture from acacia trees.

THE RAIN-SEEKERS

Although mostly sedentary by nature, blue wildebeest sometimes take part in massive migrations numbering thousands or even tens of thousands of individuals. They may follow rainclouds in a bid to track down fresh grazing and have been known to move towards the sound of thunder as far away as 25 kilometres.

Scent does not appear to play a role in their rain-finding exercises. The younger, non-territorial bulls are dispatched to the outer fringes of the herd during these migrations, where they may rely on zebras to help warn them of a predator's approach. Lions constitute their biggest threat.

Blue wildebeest in the arid reaches of northeastern Namibia. A study of wildebeest in this area shows that they spend half their lives standing, lying down or resting in the shade.

Framed by low grass and sky, blue wildebeest rest in the sun. During the rut, two or three bulls may round up harem herds numbering anything from two to 150 females with their young, sharing the herding and breeding duties.

A tiny leopard cub tries to negotiate an obstacle in its path. In the wild, females bear up to three young, but usually only one or two survive.

GAMES OF SURVIVAL

Many animals in the wild use 'play' as a means of developing survival skills. Leopard, lion and cheetah cubs learn the art of agility and speed when they dart about their mothers, rolling over her body, pawing at her nose, and pouncing on her tail. Later, in true cat-like fashion, the cubs play stalk-hunt-and-ambush games, leaping upon each other with mock aggression and comical bounds.

Such games hone the hunting skills of young animals: they exercise the muscles, develop reflexes and prepare the youngsters for stalking and catching live prey.

On the other hand, play teaches some animals how to avoid capture by improving speed and agility, and heightening awareness of a threatening presence. Young antelope, for instance, learn to bolt and swerve, young baboons learn how to flee to the safety of a tree, and young squirrels are quick to dart for cover if there's danger around. All these skills are well-rehearsed in the games animals play.

A young leopard plays with its mother's tail in the Kruger National Park. The underside of the mother's tail is white, in order to enable young cubs to follow it in the long grass. Young leopards remain with their mothers for nearly two years.

TRIVIAL PURSUITS

Nowhere in the wild is play more evident than amongst the primates. Cartwheeling, long-jumping, hurtling through the air between branches in a forest, skidding down the slopes of a donga, or playfully jumping on each other ... these are some of the antics common to monkeys and baboons.

During play, young animals use certain body signals, postures and facial expressions to show other members of their group that their antics are not serious – that aggression and flight are simulated. Such antics, often involving exaggerated body movements, put the parents of young at ease, showing them that their offspring are not actively threatening each other; that their fleeing movements are flights of fantasy and not the real thing.

A group of banded mongooses congregates around a tree in Botswana's Chobe National Park. The young forage with their parents when they are five weeks old. Banded mongooses live in troops of up to thirty.

This Cape fox launches an airborne assault during a game with one of its brothers in the Etosha National Park. Mostly shy and nocturnal, this species is the only true fox found in southern Africa.

Grooming and play form an important part of the social behaviour of vervet monkeys. Like baboons, they live in tightly knit troops of about 15-20 individuals, dominated by a mature male.

Two lion cubs in the Krugersdorp Game Reserve indulge in their favourite pastime – play. Carnivores and primates play more than most other animals.

A lioness is an attentive mother that keeps her cubs well-groomed. As they grow older, the cubs graduate from games of mock-killing to pursuing live prey brought to them by their mothers.

BODY LANGUAGE AND BITES

The games young animals play teach them a lot about the bodies of their siblings and their parents. They cement social bonds, develop muscular strength and may also influence the youngster's social standing in the eyes of its peers.

Amongst the primates and carnivores, parents spend a long time with their offspring, often showing remarkable restraint when they become the victims of their youngsters' pranks. Tail-tugging, biting and paw-slapping are common examples of the type of play indulged in by young animals.

Mostly patient and gentle with their young ones, a lioness or a baboon may cuff a youngster whose playful antics become too boisterous or irritating.

The lioness's extraordinary tolerance towards her young makes her the target of a variety of cub games. Adult males are not exempt from taunts, tugs and bites from cubs, which can, at times, be quite painful.

Copper-tinged dunes and thirstlands of vivid red

Millipedes range in size from 20 cm black giants to tiny white species that live in the perpetual darkness of caves.

An aardvark snuffles with satisfaction, fleshy muzzle buried deep in the remains of a termite mound, ribbon-like tongue lapping up scurrying ants by the hundred. It's time for dessert: powerful claws dislodge large chunks of sun-baked mud to reveal another maze of tiny passages ....

This is the Kalahari, an awesome expanse of copper-tinged dunes and flat thornveld dotted with shallow depressions that quickly fill with water after summer rains and lose it just as rapidly to evaporation and the thirsty sand. Scorning conventional borders of mountain and river, the Kalahari sweeps across southern Africa and captures almost all of Botswana, a slice of Zimbabwe's dry west, the eastern half of Namibia and part of the Cape Province north of the Orange River.

Eland, red hartebeest, springbok and gemsbok have adapted to this dry land and thrive in defiance of the odds. The bateleur hunts from the limitless reaches of an azure sky, parachuting in on live prey with wings held up and legs extended. A Cape cobra, preparing to deliver the coup de grâce *to a hapless lizard, rears up, its hood spread to reveal its livery of dark yellow.*

A blue wildebeest peers nervously over its shoulder, disturbed by a sudden movement on a nearby dune. This is enemy territory, the domain and hunting ground of spotted hyaenas, lions, leopards and cheetahs.

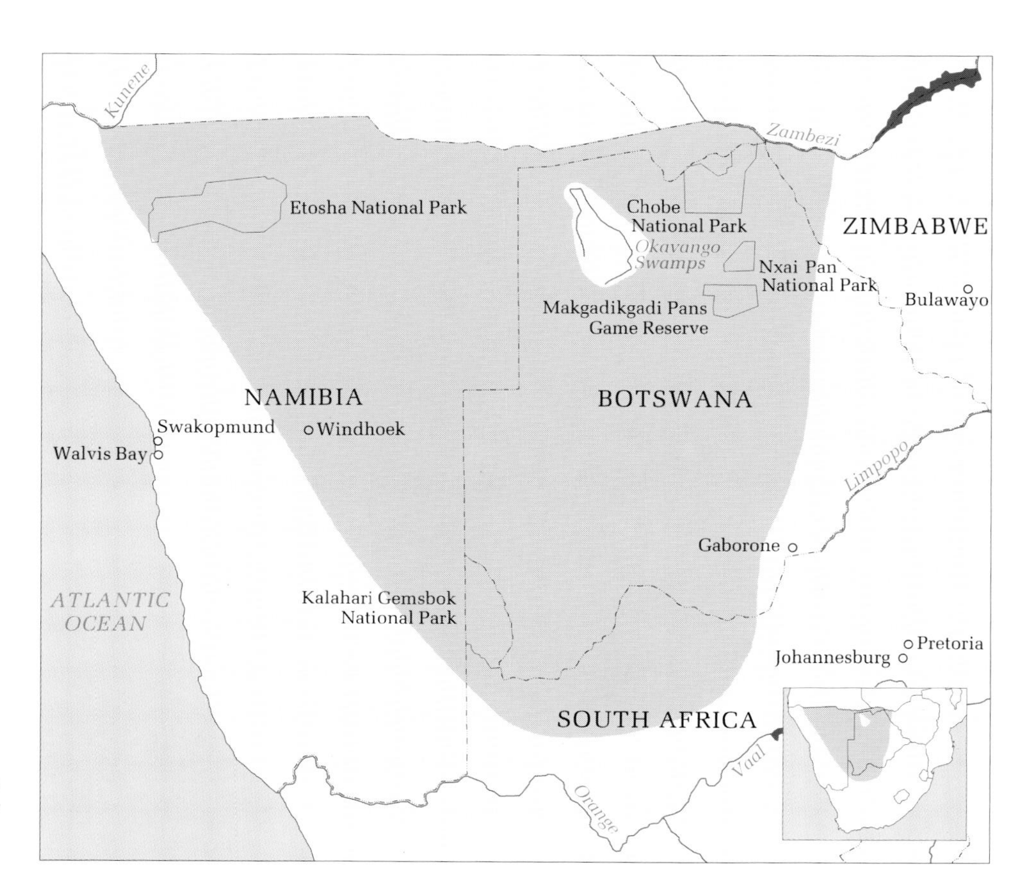

The gemsbok's rapier-like horns give it a stately air on the plains of the Kalahari. The leader of the herd, a territorial male, jealously guards his domain, marking the boundaries with piles of dung to warn off would-be intruders.

A brown hyaena scavenges for its next meal. Despite its fierce looks, it is a less efficient hunter than its spotted cousin.

In the harsh southwest, where rain is rare and the vivid red sand is sculpted into low dunes, the streams are ephemeral and most die before they reach the sea, their life sucked dry by the scorching sand.

In the moister north and northeast, elephants, buffaloes, hippos, giraffes and other large game flourish amid a bounty of thick grass, mopanes and an assortment of shrubs. At Makgadikgadi Pans to the northwest, summer rains transform the skeletal plains into reservoirs of plenty, beckoning vast herds of antelope, zebra and wildebeest.

Here and there the umbrella canopy of a camel thorn tree provides welcome shade for a gemsbok which grazes on the sweet grass, and eagerly eats the fallen seed pods of Acacia haematoxylon. *The seeds are scattered in its droppings, germinate after rain – and the survival battle begins anew.*

A baby suricate nestles in the protective embrace of its vigilant parents. Suricates will dive swiftly to their burrows at the first hint of danger.

A million red-billed queleas darken the sky as they swarm to drink at a shallow pan, landing on surrounding trees in such numbers that the branches bend almost to the ground. A crimson-breasted shrike alights on a twig, snatches a centipede and takes off before the other birds have time to take fright.

In its home beneath a rock, a Cape skink tugs vigorously at the membranous bags that envelop her newborn young. But theirs will be an all too brief lifespan: tonight a prowling honey badger will expose their sanctuary with a single slash of a knife-like claw and wipe out the entire brood.

In the southern Kalahari, a bat-eared fox lowers its head to the ground. With its huge ears held parallel to the sand like metal detectors, it listens for insect sounds beneath the surface. Having located the exact position of its subterranean prey, the fox starts digging furiously until it is rewarded with a wriggling morsel.

A honey badger or ratel's long, knife-like claws rip into the soil as it pursues a lizard. Powerfully built and formidably armed, honey badgers are extremely aggressive and can kill animals larger than themselves.

This portrait of a reclining cheetah clearly shows the characteristic 'tear marks' framing a rather disdainful nose. At rest it has a dignity and grace matched by few of the big cats.

THE GRACEFUL FLYER

The cheetah in motion is an extraordinary sight. In pursuit of prey, its high, muscular shoulders piston effortlessly over the ground in strides up to seven metres long, and its long tail is used as a balancing mechanism.

The animals move so fast that wildlife photographers experience great difficulty capturing the chase on film. Having caught and killed its prey, the cheetah feeds quickly, keeping a constant lookout for rival predators which are quite capable of driving it from its own kill.

Nervous and wary by disposition, it tends to avoid large herds of prey, preferring to attack stragglers.

Cheetah cubs view the passing show from a tree, paws bloody after gorging on a kill. The young are born in the shelter of underbrush or grass and there are usually three cubs to a litter. They start to wean at the age of five or six weeks.

Cheetahs hunt in the early morning and evening, and spend the hottest part of the day lying in the shade. They have large home ranges and although they mark out their turf, do not appear to be fiercely territorial.

A cheetah will live on the fringe of a desert, provided there's enough prey to keep it alive. Springbok and springhares form the bulk of its food in the dryer areas, although its diet also includes a variety of small mammals and even ground-birds.

FEROCIOUS FOREPLAY

Courtship among cheetahs is a curious affair. The females tend to associate with other cheetahs only when they are on heat, and approaches at any other time usually trigger an impressive display of feminine aggression. The males become very excited as the female comes into breeding condition, and show their interest with mock charges, copious spraying of urine and squabbles with other males.

A cheetah surveys its kingdom from the sanctuary of tall grass. It is equally at home in open plains and savannah woodland. Males mark out their ranges with odorous boundaries of urine and faeces.

A female cheetah with her gangly cubs, already equipped with formidable teeth. By the time they are 8-12 months old, the cubs may go hunting and even make their own kills. They quit the maternal fold singly or in groups.

This antelope didn't get away. The cheetah uses its dew claws to secure a hold on fleeing prey.

Time for a leisurely stretch. High-shouldered, spindly-legged and almost ungainly at first sight, at speed the cheetah is a symphony of grace and power. When contented, it purrs just like a domestic cat.

CUBS ON THE MOVE

Because cheetah cubs have an extremely high mortality rate – half of all cheetah cubs die before they reach three months – their mothers move their young around frequently to prevent their scent building up and attracting predators.

The youngsters are born with an 'overcoat' of long, smoky-blue hair which some say serves to camouflage them from predators in the long grass. This mantle of juvenile fur eventually disappears as the cubs mature.

The mother communicates with her young using a high-pitched, bird-like call. After nurturing and protecting her cubs for about 14 months, the mother leaves them to fend for themselves.

Alert cheetah cubs in dense undergrowth. Males form bachelor groups with strong bonds. In some groups, one cheetah always selects the prey and leads the hunt; in others they may take turns.

A cheetah's profile reveals its small, rounded head, short muzzle and high domed skull. The unusually large nasal cavities allow it to take in large quantities of air after the extreme exertions of the hunt.

The world's largest land animal has impressive dimensions, standing up to 3,5 m at the shoulder, and tipping the scales at 5 500 kg or more. Despite its bulk, the African elephant is remarkably agile and can run at speeds of up to 40 km/h.

An elephant's trunk is a highly sensitive organ controlled by more than 100 000 muscle units.

The elephant's thick skin, wrinkled like tree bark, is thickest on the back, trunk, legs and forehead, and may be 3-4 cm thick in places. A sparse layer of bristly hair covers its body, growing thickly at the ear orifices and the tip of the tail.

DANGEROUS PROTECTORS

Elephants are not normally aggressive but in some circumstances, especially if they are hurt or sick, or protecting young, they can be very dangerous indeed.

An angry elephant raises its head and trunk, extending the ears at right angles to the body and kicking up dust with its forefeet. Screaming its rage, it shakes or sways its huge head and may launch a mock charge to frighten off its enemy. But the charge is not always faked, and the elephant is quite capable of following through with a real attack.

Bulls may have fights over cows on heat, during which they may inflict serious damage with their tusks; some of these battles can end in death. Even other animals may

Young elephants remain close to their mothers, maintaining contact by grasping their tails. The mothers and other females are fiercely protective during the early years.

sometimes inadvertently incur the wrath of these normally peaceful behemoths.

Elephants are known to have killed rhinos and hippos, and occasionally have been seen chasing buffaloes, zebras and antelope from dwindling waterholes. Many people have been killed by elephants too: some were crushed with the trunk or trampled to death; others were crushed or impaled by the tusks.

Flapping ears do not, as many people believe, indicate anger. This action has the effect of cooling the blood which courses through a fine network of blood vessels just under the skin in each ear. This is Nature's way of helping to regulate the elephant's body temperature.

Elephants on the move. Very young calves actually walk beneath their mothers' bellies, where they are safe from predators. Should a nursing mother die, her calf may be cared for by other nursing females.

Elephants love water, and will spend hours bathing, lying down in it and spraying themselves with obvious pleasure. Occasionally they submerge completely, with only the tips of their trunks protruding.

An aerial view of an elephant herd. These pachyderms can move over vast ranges, and survive in habitats varying from semi-desert to plains and woodlands. Elephants prefer clean, sweet water and will travel very long distances to find it.

ANATOMY OF A GIANT

The former name of the order Proboscidea, 'Pachydermata', *refers to the elephant's thick skin, which encompasses a frame of remarkably large proportions.*

The skull within this canvas of skin houses an impressive brain, weighing between 4,5 and 5,5 kg. To cope with its diet of grasses, leaves, shrubs and soil, the elephant has a simple digestive system. The small and large intestines together may be as long as 35 metres.

The elephant's legs are positioned under its body like the legs of a table, to give plenty of support to its long vertebral column, and the heavy contents of its thoracic cavity and stomach.

Elephants do not appear to suffer from claustrophobia, and will happily rub shoulders all day, even staying close during their long hikes. Their feet have a thick layer of cartilage which works much like a shock absorber.

A pensive springbok nibbles a twig. This attractive bovid is both a grazer and a browser. It is by no means a picky eater: one study in the western Transvaal found that it dined on 68 different species of plants.

Delicate, shy and vulnerable, the Damara dik-dik shelters from the harsh summer sun. When startled, it emits a short, explosive whistle before bounding for cover in a deep thicket.

A herd of eland, largest of the African antelope. Despite their bulk (males can weigh over 700 kg, females 400 kg) they are remarkably agile and are capable of vaulting a two-metre fence with no difficulty.

A leopard spoor. Like all cats, leopards have five digits on the front feet and four on the hind feet.

Replete after a recent kill, a leopard dozes on a branch. When capturing large prey, the leopard hits it with a 'slamming embrace', delivering the killing bite at the back of the head, neck or throat.

FEROCIOUS POWER OF THE SOLITARY KILLER

Solitary, ferocious and powerful, the leopard is a superb killing machine. It will take any prey ranging from mice, birds and snakes to large antelope and even giraffes. In 1964 a leopard trapped on an island in Lake Kariba ignored impala and duiker and ate fish instead.

Leopards are stalkers and pouncers, using any cover available and their own magnificent camouflage to creep up on their targets. They hunt by day and night with equal efficiency, using their exceptional hearing and sharp eyesight to locate prey, often from a distance of several hundred metres.

When food is scarce, a leopard may make a second kill soon after feeding. In areas where other large predators occur, it lodges the carcass in the branch of a convenient tree – one account tells of a young giraffe being lifted several metres off the ground – and returns later to eat.

The same leopard, this time wide awake. Leopards generally avoid human contact. However, they are especially dangerous when hurt, cornered or suddenly disturbed.

Statuesque in a typical pose, a leopard shows why the perfect camouflage of its black spots and rosettes has been mimicked by police and military forces around the world.

Time for a spot of grooming. Its large size and wildness apart, the leopard is similar in many respects to the ordinary domestic cat, with its lazy habits and inclination to purr loudly when feeding.

Leopards appear to be independent of water sources but will drink readily wherever water is available. They can live in a wide range of habitats, from forests to rocky koppies, mountain slopes and even the fringes of deserts.

A male impala displays the ridged lyrate horns that it uses to intimidate other males in territorial clashes. This graceful antelope is a gregarious creature and gathers in herds of up to 100 during the cold, dry months.

A female impala and her calf. The single young are usually born in early summer.

Springbok, zebras and giraffes mingle without rancour at a waterhole. They move aside respectfully when elephants draw near to drink, and are especially cautious when the water supply is limited.

ROCKING HORSE RITUALS OF THE HAREM SEEKERS

Rituals of courtship and aggression are an integral part of the blue wildebeest's social organisation. It's a gregarious animal, occurring in herds of a few dozen or groups numbering thousands.

During the mating season, bulls will round up harem herds of varying size. A herd of 150 or more may be attended by one to three bulls which share the herding and breeding duties. Should their charges be approached by a bull from another herd, he will be chased off. Alternatively, the territorial bulls drop to their knees and 'fence' with the intruder, butting heads furiously but usually without doing any serious damage.

Another odd manoeuvre is the 'rocking horse' canter employed by the territorial bull to scare away rivals, which may degenerate into a head-butting contest if it fails to have the desired effect. Young males are ejected from the herd at two years.

A pale chanting goshawk wades through mud, ruffling its grey-and-white plumage. This raptor can be a ferocious predator, attacking and eating mammals up to the size of a ground squirrel, and even snakes.

Blue wildebeest on the run at Etosha. These tough denizens of the plains and open woodlands are insatiably curious and will stop and stare at an intruder for some time before whirling and running off.

The gemsbok's magnificent horns often come into play during ferocious territorial disputes between males.

A large grey mongoose warily scans its surroundings for signs of danger. Its thick, brindled coat protects it from the poison fangs of snakes, which form a part of its diet.

THE LORDS OF LANGUOR

They may be proud bearers of the title 'King of Beasts', but mostly their behaviour resembles that of lazy peasants. For surely no real monarch would allow himself to be seen sprawled on his back, feet in the air?

In fact, lions avoid exerting themselves wherever possible, and their lethargy is sometimes taken to extremes. In one instance, a male was spotted lying on his back, propped up by a tree and clearly content with his lot. Overcome by an urge he couldn't ignore, he didn't bother to get up and simply urinated in a high stream over his startled female companions.

Lion cubs spend hours in rough-and-tumble games with their siblings. Sadly, they may starve to death in areas where prey becomes scarce, because the adults consume the entire carcass, leaving little or nothing for their young.

A cub rests after feeding, his chest and forepaws still covered in the prey's body fluids. In the lion community, cubs will suckle lactating females that are not their own mothers.

Two male lions doze in the shade, reinforcing the widespread belief that it's the females who do the work. Lions are active mainly around sunrise and sunset, though they will hunt at any time of the day or night.

A lioness and her cub. Litters usually range from one to four but a litter of six has been recorded. After a kill, the youngest cubs are the last to eat, a fact which contributes to their chances of starving to death.

Can this graceless creature really be the King of Beasts? The short answer is: yes. For all of his lethargy the lion is anything but a pussycat, and his rule of the wild is fearsome.

A pride of lions at sunset. The start of the evening hunt may be preceded by elaborate rituals of yawning, mutual grooming, defecating or urinating. A roar will often trigger a chorus from the pride.

The lioness is the nucleus of the pride and the major provider of prey for her mate and their offspring. She's lighter by far than the male but her speed and power are no less impressive.

Battle-scarred and experienced, this lion has seen and done it all. The scars on his nose are probably the result of clashes with rival males, which rarely end fatally. In more serious encounters, the lion's heavy mane protects his neck from injury.

A lion shows who's the boss at a buffalo kill in the Kruger National Park. Violent arguments over food are common in a pride during hard times.

COMBAT IN THE BUSH

The endless struggle for food, the need for territory and the social hierarchies that exist amongst animal communities all contribute to conflict situations in the wild.

A hungry lion, for instance, may violently attack a lioness that interferes with his meal after a kill; many species of male antelope, including springbok and impala, will clash horns with intruders in defence of their territory; and spotted hyaenas occasionally savage their own species in the melee following a kill.

Many animals will enter into combat in defence of their young or other members of their own group. Baboons, for example, are known to attack leopards that threaten their troop in any way.

Black-backed jackals snap at each other in a furious tussle over a dead seal pup on the Namib Desert coastline. Apart from their diet of dead seals, sea birds and fish, black-backed jackals also eat mice, insects, fruit and the leftovers of predator kills.

Two Burchell's zebra stallions come to blows in Londolozi Game Reserve. Duelling rival stallions may lash viciously at each other with flailing hooves or inflict painful bites, but they never fight to the death.

STAGES OF THE FIGHT

While fatalities do occur, lethal combat amongst animals in the wild is rare. Most conflicts begin with threatening displays such as roaring, snorting, baring the teeth, pawing the ground or fluffing up hair. These displays may proceed to such physical contact as biting, kicking, slashing or horn wrestling. Duels usually end with one of the parties backing off or showing submission by crouching, rolling on the ground or sleeking the hair.

Wings splayed and feet flying, a secretary bird ruffles an opponent's feathers in a startling balletic duel as graceful as it is vicious.

Springbok males lock horns in a territorial test of strength and agility. Although these contests usually end with one of the combatants being wrestled off balance, some fights can lead to a fatal goring or a broken neck.

The Cape's ancient floral kingdom

The spotted eagle owl relies on its colouring to avoid detection. It occupies high vantage points where it can scan its surroundings for prey.

Hanging by one foot from the ceiling of a remote cave on Table Mountain, an Egyptian fruit bat twitches its sensitive ears. Then, moving faster than the eye can follow, it drops from its perch and wheels into the night, miraculously avoiding the thousands of other bats that have taken wing as if on the instruction of an unseen air-traffic controller.

The bat flutters in an erratic zigzag path to the lower slopes, its in-built radar warning of obstacles along the way as it locates a fig tree in a suburban garden. First circling the tree, it darts to a branch and buries its head in a ripe fig.

In the blackness of an underground labyrinth not far from the mountain, a Cape mole rat stirs. Silently, it slides its furry body into a smooth-walled tunnel and, using its large incisors, excavates a new passageway about 15 cm under the ground.

Although biologists rarely describe an animal as ugly, in the case of the mole rat they must be sorely tempted. With its beady, almost vestigial eyes, curiously shaped feet and bizarre dentition, it is certainly no beauty.

It's April and the Thysbe copper butterflies are at play in the fynbos kingdom, wings of copper and mother-of-pearl fluttering among the proteas, pincushions and rooibos tea bushes like scraps of windblown paper.

This butterfly enjoys a mutually satisfying relationship with a colony of cocktail ants during its caterpillar stage. The ants allow it to live with them in their underground shelter, perfectly safe from predators, while they milk the tasty sugary secretions from a 'honey gland' on its back.

Fynbos is a habitat characteristic of the winter-rainfall region of South Africa and, although it is confined mainly to the southwestern Cape, it extends eastwards for several hundred kilometres. Some 600 species of indigenous heath and erica, 70 different proteas and 50 varieties of disa flourish in this

Bontebok graze on fynbos-covered slopes near Cape Point. The males defend their territory with a grand show of head nodding, snorting, bucking and innocuous kicking. They entice females into their territory with an extravagant courtship ritual.

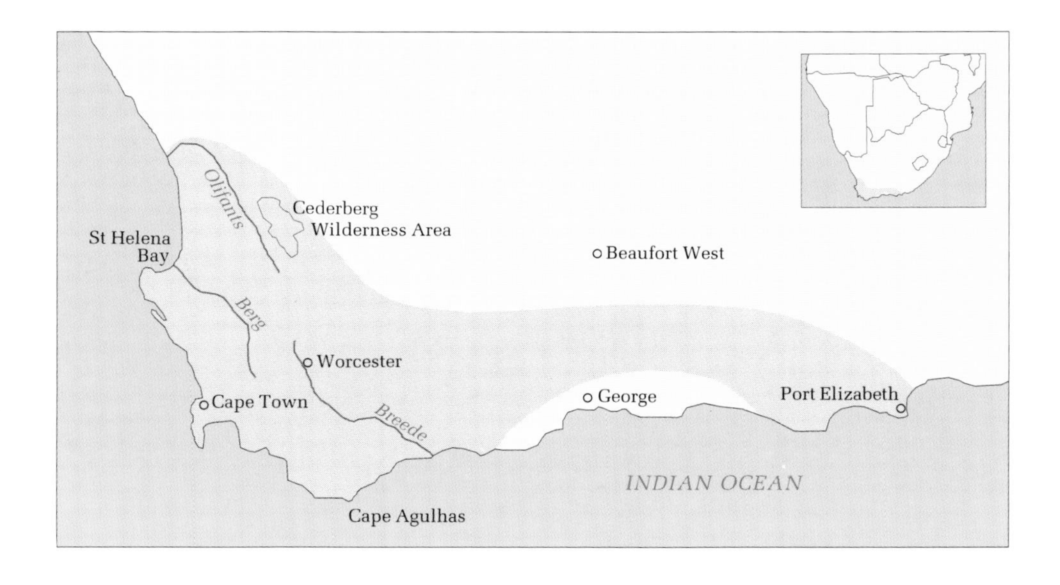

natural treasure house bounded by the Cape Fold Mountains and the Atlantic Ocean. One of six floral kingdoms, fynbos is the largest and most spectacular display of flowering plants on earth.

Known as 'macchia' to the botanist, the hardy vegetation is believed to have its origins in the misty dawn of time when the southern oceans began to cool and the cold Benguela Current off the west coast of southern Africa introduced a more arid and colder climate, with winter rains. The drier summers that followed wiped out the palm forests, chased the ancient yellowwoods to the shelter of mountain kloofs and finally, about 3,5 million years ago, spawned the macchia or fynbos that became the dominant flora of the Cape.

Elusive and quite beautiful, the rare marsh rose is one of the fynbos gems. Typically, the marsh rose starts life as a seed deposited beneath a few centimetres of decomposed sandstone on a high saddle of the Hottentots Holland mountain range. Here it is exposed to driving rain, hail and even snow before a sudden fire cuts a swathe through the surrounding bush, covering the soil with nutrient-rich ash and lingering just long enough to crack the tough seed pod.

A shoot appears, then a shrub and finally a display of blood-red flowers. Its work done, the shrub withers and disappears, the seeds bake in the summer sun – and the cycle begins again.

With his long, attractive tail feathers, the male Cape sugarbird attains a length of 43 centimetres. He perches on protea bushes, tail blowing in the wind, singing gaily, dashing after males intruding upon his territory.

In flight the European bee-eater has a clear, bubbling 'quilp' – a call that helps to keep the flock intact.

The African wild cat is an adept stalker, rounding on prey swiftly, and efficiently dispatching smaller animals with its sharp teeth.

Resplendent in its yellow, olive-green and black plumage, a bokmakierie sings to a new day. The bird is a superb vocalist.

A spotted eagle owl flies from its perch. These owls become quite vocal at night. They call in a duet, each bird giving a two-note hoot, the second note lower than the first. They are usually found in pairs that roost close to each other by day.

SILENT CREATURE OF THE NIGHT

The large-spotted genet is a cautious creature that does most of its foraging at night. During the day it holes up in a hollow tree or log, among roots, debris and rocks, or in the disused burrow of an aardvark or springhare.

Its diet includes insects and rodents (mice and rats), and wild fruit which it often eats in trees.

The normal gait of this wily animal is more of a slink than a walk, with the head held low. To gain a better view of the surrounding countryside it sometimes sits down in a 'begging' position, its outstretched tail providing balance. If alarmed, it bounds away towards cover.

A large-spotted genet peers furtively from its hideout. This genet is a very agile creature, capable of quickly scaling trees and jumping from branch to branch when hunting for food or being pursued.

THE CURIOUS STEENBOK

The delicate steenbok has many enemies, ranging from pythons to wild dogs, leopards to martial eagles. But this small, slender antelope can put on a fair turn of speed when threatened. However, it usually opts for discretion before flight and when frightened will lie down in the grass, ears flattened against its head, or seek refuge in an abandoned aardvark hole.

If flushed from its hiding place, the steenbok jumps to its feet and darts off, pausing after some distance to glance back at its pursuer – a curious habit that may have fatal consequences. Male and female steenbok do share territories, but do not necessarily associate closely together.

Steenbok live in open country where grass is abundant and the vegetation provides sufficient cover for their defensive needs. Their home turf has clearly defined areas for resting, feeding and urination.

The steenbok marks its home territory with secretions from glands in its feet, beneath its eyes and on its throat.

A bushbuck grazes in a forest clearing. These shy animals have acute senses of sight and smell. Threatened by a hostile presence, both sexes may issue a hoarse warning bark, before bounding off into the cover of a thicket.

When disturbed, the klipspringer emits a shrill alarm cry, which the ewe echoes.

The female red hartebeest leaves the herd in summer to give birth in a sheltered place. After birth, she licks up her calf's urine and faeces to eliminate telltale odours that might attract predators.

BALANCING ON PINNACLES OF ROCK

The klipspringer has long, narrow hooves, but treads only on the rounded tips – like a ballerina. The front tip is hard and rubbery, steadying the antelope as it absorbs the shock while bounding from rock to rock. The animal can balance skilfully, with all four feet held together, on the pinnacle of a rock whose surface is no more than 5 cm in diameter.

Hollow, stiff bristles give the klipspringer's coat a springy texture that protects its body from bruising contact with sharp rock edges. The coat also acts as an insulator, conserving body moisture.

The common duiker feeds at dawn and dusk, with its main period of browsing well after sunset. Weighing about 20 kg, this little animal rarely eats grass, but prefers fruit and the seeds of trees.

A young leopard lies next to its mother. Females rarely manage to rear more than two cubs per litter due to the scarcity of food and the threat of predators.

One of the most streamlined, efficient hunters in the wild, the leopard will take any warm-blooded prey, from mice to mammals twice its own mass. A leopard has been known to drag a 100 kg camel carcass into a tree.

Leopards defend their 20 sq km territory all year round, marking the boundaries by urinating and scraping tree trunks. Within this demarcated area the leopard travels tirelessly in search of food – as far as 25 km in one night.

LEADERS OF THE TROOP

In baboon society, it is the males who are firmly in charge. This dominance is drilled into the baboons from an early age, and although chacma babies are reared with devotion, any disobedience or disrespect will earn a swift cuff from an admonishing male. Although a showdown with another troop involves more noise and spectacle than anything else, don't write off chacma males as all bombast and show.

With their powerful build and large canines, they can fight viciously to the death, and old males can become treacherously bad-tempered.

Chacma baboons spend hours grooming each other, locating and catching fleas. Social bonds are very strong in baboon troops, with a hierarchy of dominant males maintaining the pecking order.

An infant chacma baboon cautiously surveys its environment. Its mother will carry the baby around for a few months, grooming it constantly, and will not allow other females to pick it up until it can walk.

OASES OF DEATH

Fountains of life, oases in the desert, cosmetic parlours for sun-baked mammals ... waterholes are all of these. But a sinister, more deadly, theme underlies the existence of these muddy pools of sustenance and survival: for many animals in the wild they are no more than the waters of death.

In the dry season, when plains animals such as springbok, impala, wildebeest and zebra converge on waterholes to drink, the large predators follow them, assured of a plentiful supply of food.

For lions, a waterhole is the ideal place for a surprise ambush – it not only saves the lion energy, but it increases its chances of making a successful kill. Driven by thirst, many animals, including kudu, giraffe, warthogs and even young rhino cast caution aside and wander down to waterholes, only to find a hungry lion lying in wait.

A waterhole brings cool relief to this group of buffalo at Londolozi in the eastern Transvaal. Buffalo are primarily grazers, and like to drink twice a day.

A group of Burchell's zebra cools off at the Klein Namutoni waterhole in Namibia's Etosha National Park, while a lone kudu bull looks on. Timid and restless, these zebras are quick to bolt from a waterhole after drinking.

A white rhino takes a mud bath at a waterhole in Hluhluwe Game Reserve. A grazer, not a browser, it uses its wide, flat mouth to crop short grasses.

After quenching its thirst, an elephant takes a muddy shower at a waterhole. An elephant sprays vast amounts of water over its body daily to keep cool.

Dusting down is a popular alternative to mud-bathing, and forms an important part of daily hygiene in elephant society. The mud or dust creates a coating which protects the elephant's sensitive skin from biting flies.

Red hartebeest and zebra share a waterhole in Namibia's Etosha National Park. Antelope are particularly vulnerable to predation at waterholes.

THE MUD WALLOWERS

While waterholes serve their purpose as providers of liquid sustenance, they have another important cosmetic function for elephants and rhinoceroses. These animals immerse themselves in mud, allow it to dry, and then shake it off, a procedure which eliminates most of the parasites clinging to their hides.

Young elephants find waterholes ideal for playing, wrestling and splashing one another, and you'll often see them wallowing in the mud under the watchful gaze of their parents.

A black (hook-lipped) rhino cools down in a Zululand mud wallow. The prehensile upper lip of this species is used to grasp the twigs and shoots on which it browses.

Haunting cries and logs with eyes

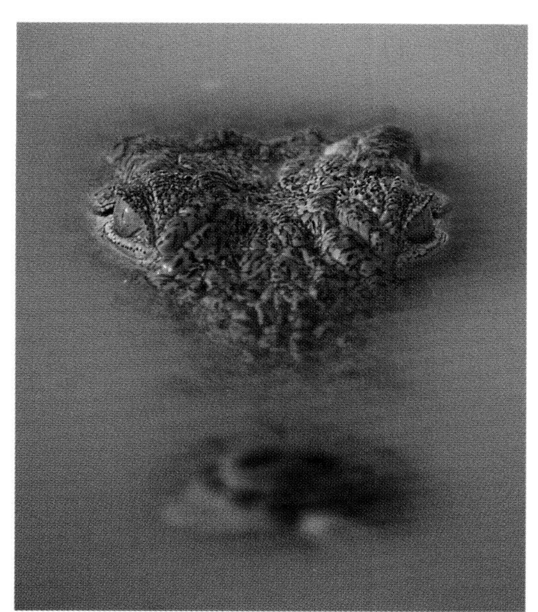

Barely noticed by the passing parade and by potential prey is this almost submerged, lazily relaxed but ever alert Nile crocodile.

A haunting cry echoes across the vast, shallow waters of the Okavango Delta, momentarily stilling the ragged chorus of chirps and croaks from the reed-covered islands. On a treetop at the edge of the swamp, an African fish eagle poises for flight, eyes fixed on a dark shape moving just beneath the surface of the still, limpid water.

It drops into a long, shallow glide, strikes the water with a barely noticeable splash and soars upwards, a wriggling bream clutched firmly in its talons. There's a glimpse of the eagle's maroon belly and underwings as it returns to its perch.

The show is over. The frogs resume their song, spindly-legged pond-skaters continue their erratic dance, and the waters of the Okavango meander into oblivion as they have done since time immemorial.

One chorister is missing, though. A green water snake has snatched a frog from its water lily 'stage' and is swimming back to the river bank where it will consume its prey at leisure.

In the labyrinth of misty waterways, cold-eyed Nile crocodiles bask open-mouthed in the slanting rays of the late afternoon sun. One slides into the water as if on rails, scaly body submerging until only its eyes and nostrils protrude from the surface. Then it disappears into the murky depths, a valve shutting in its throat to keep out the water: soon a nesting duck will die with a single snap of those terrible jaws.

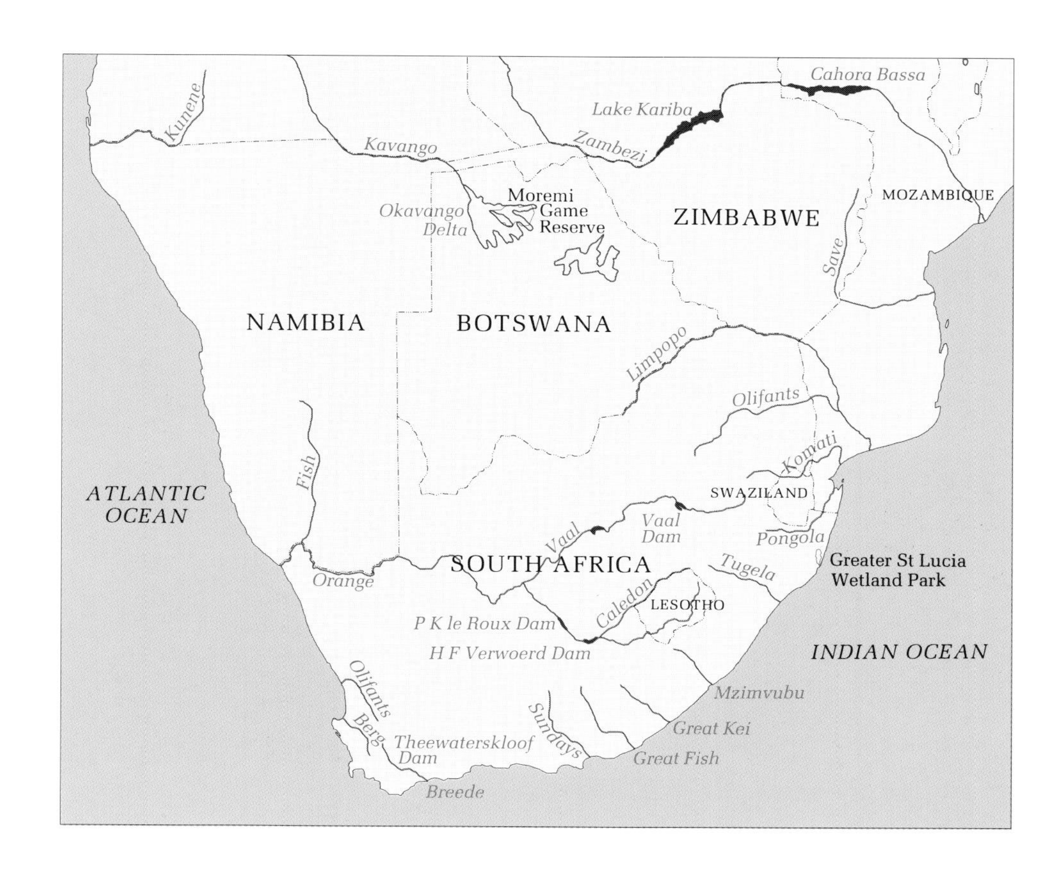

A common but always fascinating sight along the meandering waterways of Botswana's Okavango Delta is the red lechwe, a cousin of the waterbuck. Clumsy on land, lechwe are superbly graceful swimmers.

The distinctive little pied kingfisher, often seen hovering over water but here at rest. The single broken band across its upper chest identifies it as a female.

Nearby, between floating beds of papyrus, a scrum of hippos huff and snort in silent channels. Grumpy and unashamedly somnolent by nature, their voracious appetites will take them long distances in a single night in their search for grass and the young shoots of reeds. Sleepy as they may seem, hippos make formidable adversaries and even crocodiles will treat them with the utmost respect.

Deep in the tangled vegetation that lines the waterways, insects move unseen in a dim, green world. And myriad species of birds, their flight hastened by the imminent sunset, head for their treetop or waterside retreats.

Such is life in the freshwater wetlands of southern Africa.

Down in St Lucia Estuary on the Natal coast, the battle for life is ruthless: millions of microscopic organisms pounce on waterborne nutrients and are quickly gobbled up by hungry fish. The fish, in turn, flee the predations of a host of amphibians,

White flank stripes and a wheezy two-part whistling call distinguish the fulvous duck, which loafs in large groups on the fringes of inland waterways.

and large birds that dive in on them or pluck them from the water with long, sharp bills.

This is the kingdom of mud bream, barbel and grunter; of gulls, terns, plovers, sandpipers, kingfishers and many other feathered marvels. A stork paces the shallows and pauses to contemplate its lot – standing on one leg. A hundred flamingoes take to the sky, setting it alight in a long skein of pink. A pair of seagulls perch on the bank beside a half-eaten fish, squabbling over the remains and exchanging curses in a dialect that only gulls can understand.

As evening falls, there is movement everywhere. A water mongoose jabs at a hole in the bank with its forelegs, long claws gouging away chunks of clay until an estuarine crab is uncovered. It crunches the carapace with its powerful teeth and swallows the contents before swimming to the opposite bank in search of frogs ....

A darter stands poised on a branch, ready to spear a fish with its bill.

Well-developed curved claws enable the Nile monitor to surmount high obstacles, and even to climb trees, but water is its preferred habitat. These lizards grow to two metres in length, and have a fondness for crocodile eggs.

MONSTER OF THE WATERWAYS

Huge, heavily armoured and probable source of the biblical leviathan, is the Nile crocodile, nearly six metres long and feared throughout the African continent as a man-eater. And indeed, humankind does feature tragically often in the giant amphibian's diet, though its usual prey comprises a variety of fish, waterbirds and antelope, the larger of which it snatches from the water's edge and kills by drowning.

Also awesome-looking but in fact harmless enough – to man at least – is the Nile monitor, or water leguaan. This enormous, strikingly coloured lizard will arch its back, thrash its powerful tail and hiss menacingly when threatened, and then slip quietly away or, if cornered, feign death.

The Nile crocodile is the only member of the order Crocodilia *indigenous to southern Africa. The massive reptile is at its most visible on the sunlit sand banks of rivers, where it basks in order to gain body heat.*

Greater flamingoes skim the flat plains of Botswana in their thousands. The bright red patches on their forewings are most noticeable in flight.

Ponderous on the ground, noble in flight: the white pelican's three-metre wingspan carries it through the air with graceful ease. Few sights are more entrancing than a soaring, V-shaped formation passing overhead.

THE GRACEFUL KILLER

Few who see it in action or who hear its loud, hauntingly plaintive cry across the waters can ignore or forget the African fish eagle. The bird, one of the continent's larger raptors (it has a two-metre wingspan) is a masterly and quite spectacular hunter.

It swoops down at a shallow angle, arrowlike, to clutch its prey, usually a surface-living fish, in powerful talons, before bearing it away to eat at leisure. A dramatic sequence indeed, but the image tends to mislead: fish eagles actually spend very little of their time on the hunt (one Kenyan study puts the figure at just eight minutes a day). Nor is their diet restricted to live fish – they often resort to robbing other birds of their eggs and nestlings, and have even been known to scavenge.

A swooping fish eagle descends on its prey. These birds are real home-lovers; in many instances a pair will spend its entire lifespan in the same small patch of wetland.

Gigantic jaws and a two-ton body weight make the hippo a formidable foe. Fights between adult males can lead to the infliction of gaping wounds.

The hippopotamus' name, aptly enough, is derived from the Greek word for 'river horse'. These heavyweight amphibious mammals boisterously congregate in groups of between 10 and 20 individuals.

Hippos can stay submerged for up to five minutes, during which time they may swim or walk along the river bottom. They prefer sources of water where they can stand with their heads or upper bodies above the waterline.

Insects and frogs feature prominently in the diet of the hinged terrapin, one of only two southern African freshwater tortoises.

The most colourful of southern Africa's amphibians are the frogs of the genus Hyperolius. *Common names of the 16 species, such as arum frog and water lily frog, pinpoint their various habitats. This is the reed or rush frog.*

This woolly coated water mongoose is a swift and efficient hunter – it has just gouged a crab, one of its favourite meals, from the river bank. It also eats frogs, rats, mice and insects.

A young waterbuck, threatened by a leopard, seeks the safety of the river. The two small objects in the water on the right are terrapins.

Waterbuck are grazers, though occasionally they will eat leaves and even certain fruits. They associate, invariably near water, in small herds of between 10 and 30 individuals.

The appealingly placid, moist-eyed face of the female waterbuck. Her single calf, born in summer, will spend its first vulnerable month in a well-disguised hiding place.

Waterbuck are among the larger, prouder looking antelope: males stand 1,3 m at the shoulder; the needle-sharp horns are about 90 cm long, and fights between competing bulls sometimes end with the death of a combatant.

SHAGGY COATS AMONG THE REEDBEDS

One of the few African antelope to escape the attentions of the human hunter is the dewey-eyed waterbuck. Its protection: an oily substance, produced by glands on the skin, that has a highly unpleasant musky smell, and which taints the flesh. Against other predators, however, the animal uses more conventional defences.

When threatened – as it is by leopard, cheetah and, especially, by lion – it will retreat to the deeper water of the reedbeds, submerge itself to the neck and face its pursuer. Male waterbuck are taller, heavier and rather more handsome than the females – an adult bull weighs in at 270 kg, and boasts a superb, forward-curving pair of horns, which it uses with bloody effect in its fights over both territory and possession of the nursery herd.

Lack of horns, a smaller build and a shorter, shaggier neck distinguish the female from the male waterbuck. Both sexes, though, display the highly visible white ring that encircles the rump.

Quest for food on the ocean's fringe

A simple scrape in the sand is home to this white-fronted plover chick. It eats tiny marine creatures on the shore.

The Cape fur seal presents a comical, somewhat ungainly appearance on dry land as it waddles on bent flippers like a tipsy old gentleman, moustache bristling above its white 'bib'. But once in the water, it becomes a glistening torpedo, a symphony of grace and speed.

Diving deep beneath the waves, the seal streaks after a passing shoal of anchovies, swallows one whole and chases another, then another. Within minutes it has dispatched several fish – enough for now. Swimming towards an island, it clambers up a rock, using the 'prints' on the soles of its front flippers to grip the slippery surface while it hauls itself out of the water.

It's early November, and the breeding season has just begun. The fiercely territorial males have established their terrain even before the females come ashore and mating takes place. Soon the rocks will be a maelstrom of furry pups, vulnerable and totally dependent on their mothers for food and protection.

From a nearby dune, a black-backed jackal watches intently as the seals bicker and doze in the late afternoon sunshine. Its gaze flickers repeatedly to the carcass of a long-dead female at the edge of the surf: the jackal has survived on tiny scraps of carrion for two weeks, and the pangs of hunger are making it increasingly desperate. Later it will slink down among the sleeping colony and feast on the remains of the seal.

This is life on the southern African shoreline, the coastal fringe that runs along the border of the Namib Desert's dune

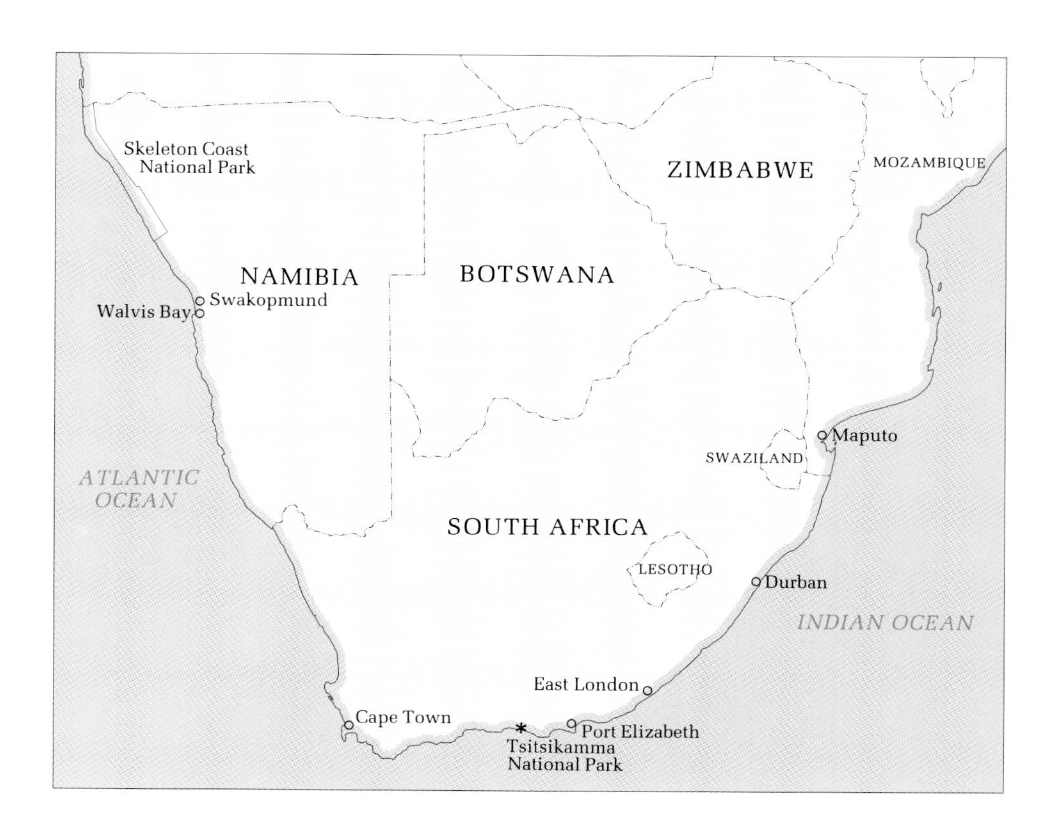

A curious and common sight on the desolate Skeleton Coast of Namibia is the black-backed jackal. Seal pups, birds' eggs and carrion are important ingredients in its overall diet.

Vast colonies of black Cape cormorants darken the offshore islands from Cape Agulhas to Namibia.

seas, down South Africa's west coast to the Cape of Good Hope and up to the subtropical splendour of Natal.

Offshore along this coastline, particularly along the west coast, the nutrient-rich waters teem with millions of creatures, from tiny microscopic organisms to the southern right whales that cruise the coastline in late winter, spring and early summer.

Currents play a crucial role in the life cycles of these animals, determining the density and location of food resources and shifting entire populations. The cold Benguela Current flows from south to north on the Atlantic coast, the warm Agulhas Current flows southwards down the Indian Ocean coast and the West Wind Drift flows from west to east. Other strong influences in this remorseless shift are the depth of the sea and contours of the continental shelf.

One of the most formidable of the denizens of the deep is the killer whale. Weighing as much as eight tons, and equipped with 40-50 teeth in a wide mouth, the killer whale surges through the sea at 40 km/h, hunting down fish, penguins, seals, turtles and even other whales with a ferocity that has to be seen to be believed. A sophisticated echolocation system enables it to locate its prey and communicate with others of its species.

Like other mammals, the female killer whale suckles her young; the teats are located in a pocket, enabling her offspring to feed without the intrusion of seawater.

Less fierce but equally attractive, the common dolphin ranges in warmer inshore and offshore waters all along the coast of southern Africa, but is seen mostly off the south and east coast, swimming swiftly and often acrobatically in schools of anything between 20 and 200 animals. Dolphins observe a strict hierarchy in these schools, the males dominating the females, and each gender observing its own ranking order. Averaging 2,1 m in length, they are among the most striking of all the marine mammals with their black or dark brown backs, white bellies and buff or grey marking on the flanks.

Some sea creatures, like the long-spined limpet, are content with a lower profile. Occurring on rocky shores between False Bay and Cape Vidal to the north, this limpet cultivates its favourite brands of algae and actually defends its seaweed 'garden' from interlopers. It has even been seen thrusting its spines under the shell of an intruding limpet to drive it away.

Common terns skim and wheel above the waters in dense, feathery clouds. The seabirds, among the most prolific of the coastal regions, breed in the northern hemisphere, migrating south for the summer.

Coastal pollution and the depredations of man have drastically reduced the jackass penguin population, which now numbers some 130 000.

Jackass penguins preen in unison. The birds are so named for their loud, braying call, part of the courtship ritual. The birds roost in colonies on offshore islands, spending their daytime hours foraging at sea.

GIANTS OF THE DEEP CALL IN TO CALVE

If you go down to the Cape south coast's Walker Bay on a blustery winter's day, the chances are you'll see the dark and massive bulk of a whale breaking the surface of the choppy waters. This is the southern right (so called by the early whalers because it was considered the 'right', or most profitable, species to catch) that comes inshore at various points to give birth to its young.

These once-numerous marine mammals are the world's second-largest living creatures after the blue whale. An adult reaches 18 m in length and weighs around 60 tons. They have been hunted to the point of scarcity for their oil but are now protected by international convention.

A southern right whale close inshore. These gentle, krill-eating giants are most often seen during the calving season (June and July) along the stretch that runs from Plettenberg Bay to Hermanus on the Cape's south coast.

A young Cape fur seal frolics in the foam. These attractive fish-eating mammals spend about a third of their lives on dry land.

Parental discipline in Namibia's Cape Cross Reserve, sanctuary for 80 000 fur seals. The males begin establishing their territories in the latter half of October; the first pups appear about a month later.

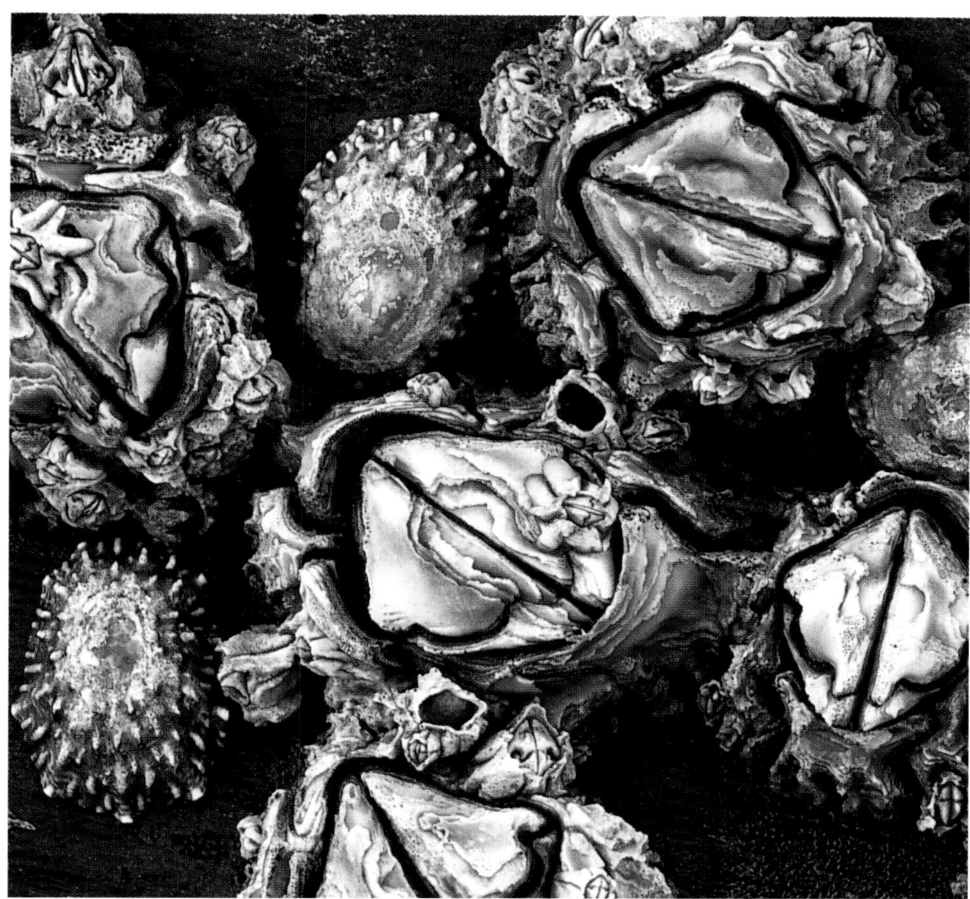

Barnacles and jagged-edged limpets cluster on a seaside rock. Remarkably, one type of limpet will actually guard its own 'garden' of favoured seaweed.

Dolphins, reputed to be amongst the most intelligent and sociable members of the wild kingdom, appear to like human company and have their own remarkably intricate 'language'.

One of southern Africa's more enterprising (and noisily aggressive) seabirds is the kelp gull. Mainly a scavenger, it will also catch live shellfish.

Airborne gannets on the lookout for pilchards, their favourite food. The hunt is spectacular: once they locate a shoal, the birds will dive en masse to pick off the fish with pinpoint accuracy.

Though their numbers have declined with the commercial overfishing of pilchards, crowded colonies of Cape gannets still roost and nest on the islands off the west coast. Their droppings, called guano, are a valuable source of fertilizer.

A lion cub stands on its mother's reclining body in Tanzania's Serengeti National Park. Cubs left unattended while their mother is out on the hunt can fall victim to such animals as leopards, hyaenas and wild dogs.

THE DRIVE TO SURVIVE

Survival is the overwhelming challenge facing young animals in the wild. Even lion cubs are at risk of dying prematurely – of disease, starvation and predation by other animals, including other lions. The adults of most species are uniquely adapted to teach their offspring how to survive by actively helping them or by example.

Lions and cheetahs teach their young how to hunt by presenting them with a quarry that is easily caught. Young vervet monkeys and baboons learn the art of foraging for fruits, berries, flowers, seeds and insects from the example of their parents, and a banded mongoose that stumbles across a rich source of food, will summon the young to join the feast.

A young vervet monkey rests between bouts of foraging. The bonds between females and their young are very strong. Females will even adopt young from other troops, and care for them as their own.

A female wildebeest lavishes affection on her newborn calf at Etosha Pan. The young are born in early summer, to take advantage of the summer rains.

Young warthogs forage close to their mother. When running for shelter, the youngsters scamper into their burrow head first, but adults do a remarkable about-turn at the entrance and reverse in.

THE BONDS OF PARENTHOOD

In the animal world, the bonds between adults and their young assert themselves most strongly in times of crisis – when an offspring's life is threatened – or shortly after birth, when a young animal is quite incapable of fending for itself.

A herd of wildebeest may form a protective circle around a calf threatened by a predator, and even charge down on the unwanted presence. Similarly, Burchell's zebra stallions may use bared teeth and flailing hooves to attack and maul a hyaena trying to round on a foal. Female warthogs, observing a threat to their offspring, may also cast discretion to the wind, and pursue a hungry predator with bravery and determination.

Elephant society is particularly interesting for the remarkable bonds that exist between parents and their offspring. Cows have been known to carry the bodies of dead calves around for days. During the first few months of its life, a calf will seldom stray more than a few metres from its mother's side.

A young suricate is dwarfed by its mother near their burrow in the Kalahari. Suricates constantly survey the sky for lethal raptors.

A newborn springbok calf stands precariously as its mother bends protectively over it. The female hides her calf in bush or long grass for the first two days of its life.

A lioness pads through the grass, carrying her young cub by the scruff of the neck. The cubs, numbering from one to four, are born in grass thickets, reed beds or clumps of bush where they are well hidden from other carnivores.

A female chacma baboon shows the strong maternal bonds that exist between her and her offspring. A high degree of dependence keeps mother and infant in close contact in the early months of a young baboon's life.

An elephant calf flanked by older members of the herd. The calf is born away from the group, and starts to suckle as soon as it can stand.

GROWING UP

The dependence of an animal's offspring on its parents varies considerably amongst different species: springbok and tsessebe calves develop so fast that within a few days they are strong enough to join the herd. Here they tend to form little nursery groups of their own which are supervised by one or more ewes.

Infant chacma baboons, on the other hand, are very dependent on their mothers, hanging to the underbelly, and later riding astride the mother's back.

When it comes to venturing out into the great wide world, otters try to encourage their youngsters to jump into the water by showing them food in the water.

An adult female baboon and two youngsters catch the final rays of a dying day. Baboons usually seek shelter for the night well before sunset.

A lioness relaxes with her cubs. The cubs stay with their mother for about two years, during which time she teaches them how to hunt by bringing them half-killed small animals which they learn to pursue and consume.

CAMEROON
CENTRAL AFRICAN REPUBLIC
SUDAN
ETHIOPIA
GABON
CONGO
ZAIRE
UGANDA
KENYA
RWANDA
BURUNDI
TANZANIA
Masai Mara National Reserve
Tsavo National Parks – East and West
Serengeti National Park
ANGOLA
ZAMBIA
MALAWI
Kafue National Park
Mana Pools National Park
MOZAMBIQUE
Skeleton Coast National Park
Etosha National Park
Chobe National Park
Moremi Wildlife Reserve
Nxai Pan National Park
Hwange National Park
ZIMBABWE
Makgadikgadi Pans Game Reserve
Gonarezhou National Park
INDIAN OCEAN
NAMIBIA
BOTSWANA
Namib-Naukluft Park
Blyde River Canyon Nature Reserve
Kruger National Park
Timbavati Private Nature Reserve
Sabi Sand Game Reserve
Kalahari Gemsbok National Park
ATLANTIC OCEAN
Pilanesberg National Park
SWAZILAND
Ndumo Game Reserve
Itala Game Reserve
Mkuzi Game Reserve
Hluhluwe Game Reserve
Greater St Lucia Wetland Park
Royal Natal National Park
Umfolozi Game Reserve
SOUTH AFRICA
LESOTHO
Cederberg Wilderness Area
Addo Elephant National Park
Tsitsikamma National Park

Celebrated sanctuaries of the southern African wild

Southern Africa has more than 700 wildlife havens, ranging from tiny reserves to vast national parks. This A-Z gazetteer provides a practical introduction to the subcontinent's more interesting sanctuaries. Each entry gives the location of the reserve, explaining how to get there and what species of wildlife you are most likely to spot.

Included are details on climatic conditions, the best time to visit, opening and closing times, how to get around inside the reserve, accommodation and booking. The text also spells out special precautions to take during a visit to a reserve. This ranges from advice about local drinking water, to taking anti-malaria tablets or administering sun protection lotion.

The parks included in this gazetteer are popular for their diversity of animals, the natural splendour of their surroundings and the amenities they offer to nature lovers.

There is the starkly beautiful Blyde River Canyon Nature Reserve in the Transvaal, with its baboons, vervet monkeys, leopards and bushbuck; Chobe National Park in northern Botswana, with its searingly hot summers, rugged terrain and wealth of large mammals; Etosha National Park in northern Namibia, with its vast herds of plains animals and prolific birdlife; Hluhluwe Game Reserve north of Durban, the Kalahari Gemsbok National Park in the northwestern corner of South Africa, Kenya's Masai Mara National Reserve, private game reserves of sheer hedonistic splendour – and many others.

Among the most celebrated reserves is South Africa's Kruger National Park, a 19 615 sq km slice of the northeastern Transvaal bushveld that accommodates elephants, white and black rhinos, hippos, lions, leopards, buffaloes and myriad other creatures large and small.

Addo Elephant National Park

Location: 72 km north of Port Elizabeth. Map page 256, reference: E14.

Wildlife: Elephant, black rhino, buffalo, eland, kudu, hartebeest, bushbuck, grysbok and duiker.

Climate: Warm to temperate, with some rain throughout the year.

When to go: Throughout the year.

How to get there: From Port Elizabeth, take the N2 towards Grahamstown. About 10 km outside Port Elizabeth, turn left to Addo. The turnoff to the park is on your right, some 15 km after Addo.

Opening time: From 15 minutes after sunrise to 15 minutes before sunset every day. From March to April the gates open at 6.30 am and close at 6 pm; between May and August from 7 am to 5.30 pm; between September and February from 6 am to 7 pm.

Getting around: Visitors are provided with a map showing the park's 43 km of roads and the numbered orientation points. No motorcycles, caravans or open cars are allowed in the park. Petrol is available during the normal trading hours, but not diesel fuel or bottled gas.

Special precautions: No citrus fruits may be taken into the park. (A previous, well-intentioned but misguided ritual of feeding oranges to the elephants made them addicted to the fruit and they became aggressive as they competed for it. Also, as they would not risk leaving the feeding area, the immediate vegetation bore the full brunt of the rest of their feeding, and plant densities were drastically reduced by over-grazing.) No pets are allowed, and all firearms must be sealed at the office.

Accommodation: Addo has six rondavels, 24 three-bedded chalets and two guest cottages for visitors. All are self-contained with electricity and refrigerators. The rondavels are provided with a communal kitchen, with crockery and cutlery. There are 20 caravan and camping sites with electricity, fireplaces, ablution blocks with hot and cold showers, and laundry facilities. There are pleasant picnic sites for day visitors.

Booking information: You can book up to 12 months in advance by contacting the National Parks Board, P O Box 787, Pretoria 0001 (telephone Pretoria 3431991); or the National Parks Board, P O Box 7400, Roggebaai 8012 (telephone Cape Town 222810); or the National Parks Board, P O Box 774, George 6530 (telephone George 746924). For more information about the park, contact the Park Warden, Addo Elephant National Park, P O Box 52, Addo 6105 (telephone Addo 400556).

Blyde River Canyon Nature Reserve

Location: Eastern Transvaal Lowveld, about 350 km north-east of Pretoria. Map page 256, reference: G11.

Wildlife: The breathtaking scenery is home to four of southern Africa's five primates: chacma baboon, vervet and samango monkey and bushbaby. You can also see bushbuck, bushpig, mountain reedbuck, klipspringer and leopard. The rich birdlife includes a breeding colony of the rare bald ibis.

Climate: Warm summers (can be very hot at Swadini) and cold winters with frost.

When to go: The reserve and its two resorts, Overvaal Sybrand van Niekerk resort, on the northern side of the river, and Overvaal Blydepoort on the southwestern side, are open throughout the year. The resorts tend to be rather crowded at weekends and in school holidays.

How to get there: From Lydenburg, follow the tarred road north through Ohrigstad (42 km), and one kilometre north of the Echo Caves turnoff, take the road to the Blydepoort Public Resort (88 km). From Hoedspruit follow the R527 for 16 km, turn onto the R531, and then take the turnoff to Swadini after about 19 km.

Opening time: Open 24 hours a day.

Getting around: Motoring in the reserve is severely restricted through lack of roads, though the various viewpoints are easy to reach. Lookout points like God's Window and Lowveld View are readily accessible from the public road running along the reserve's western edge. Hikers, however, can move more freely. There are short nature walks around Blydepoort resort and Swadini's visitor centre, while longer hiking trails lead throughout the reserve.

Special precaution: Take a course of anti-malaria tablets before visiting the area.

Accommodation: The Overvaal Blydepoort resort offers fully equipped chalets with up to five beds in two bedrooms, a sitting room, bathroom and kitchen. There is a caravan and camping site. The Overvaal Sybrand van Niekerk resort at Swadini provides family chalets with six beds in two bedrooms, sitting-cum-dining room, bathroom, fully equipped kitchen and patio. Again there is a caravan and camping site.

Booking information: For information on the reserve as a whole, write to the Officer-in-Charge, Blyde River Canyon Nature Reserve, P O Box 281, Hoedspruit 1380 (telephone Hoedspruit 35141). Addresses for the two resorts are: The General Manager, Overvaal Blydepoort, Private Bag 368, Ohrigstad 1122 (telephone Ohrigstad 80155/80158); and the General Manager, Overvaal Sybrand van Niekerk, P O Box 281, Hoedspruit 1380 (telephone Hoedspruit 35141).

Cederberg Wilderness Area

Location: Western Cape, between Citrusdal and Clanwilliam. Map page 256, reference: C13.

Wildlife: Klipspringer, baboon, steenbok, bat-eared fox and Cape leopard.

Climate: Hot, dry summers (temperatures up to 35°C); cool, wet winters, with snow on the higher peaks.

When to go: Summer is the best time to visit.

How to get there: Take the N7 from Cape Town to Clanwilliam. Halfway between Citrusdal and Clanwilliam turn right to Algeria. From the turnoff, a gravel road crosses the Nieuwoudt Pass to reach the Algeria Forest Station.

Opening time: 8 am to 4.30 pm, Monday to Friday.

Getting around: Few roads cross the area, and most visitors explore by following the hiking trails.

Special precautions: Take your own map (available from the Chief Nature Conservator's office at Algeria), food, cooking equipment, sleeping bags, tents and water (in summer).

Accommodation: There are camping sites on the Pakhuis Pass to the north and at Algeria on the Rondegat River. Two self-catering chalets on the Uitkyk Pass accommodate 22 people. The Nieuwoudts' private farms also offer camping sites and self-catering chalets.

Booking information: Campsites and hiking facilities should be reserved three months in advance. Contact the Chief Nature Conservator, Private Bag X1, Citrusdal 7340 (telephone Clanwilliam 3440) during office hours. As most of the trails cross both state and private land, permits will also be needed from the private owners. Contact either the Nieuwoudt brothers at Dwarsrivier, P O Cederberg 8136 (telephone Clanwilliam 1521), or A P C Nieuwoudt and Son, Krom River, P O Citrusdal 7340 (telephone Clanwilliam 1404).

Chobe National Park

Location: Northern Botswana, 13 km south of Kasane. Map page 256, reference: E9.

Wildlife: Elephant, buffalo, zebra, lion, hyaena and a wide variety of antelope.

Climate: Hot summers (up to 38°C), with heavy rains between November and May. Winters temperate.

When to go: Throughout the year, except for the Savuti Channel and Mababe Depression area which are closed between December and March – the rainy season.

How to get there: Take the 69 km tarred road from Victoria Falls in Zimbabwe to the border post at Kazungula. A further 13 km takes you to the northern tip of the park at Kasane. There are light aircraft landing strips at Kasane, Savuti and Linyanti (west of the park), and a new international airport at Kasane.

Opening time: 6 am to 6 pm.

Getting around: Roads are accessible for normal saloon cars only along Chobe River front; elsewhere a four-wheel drive vehicle is a must. Contact the authorities to check on the state of the roads.

Special precautions: Take a course of anti-malaria tablets before visiting Chobe. Also take mosquito repellent, a sunhat and warm clothing for the evenings, which can be bitterly cold. No swimming is allowed in the park's waters.

Accommodation: Savuti and Linyanti have luxury tented camps. The Chobe Safari Lodge at Kasane has chalets and camping facilities. Savuti, Serondela and Nogatsu offer public camping sites. Kasane also offers two luxury hotels.

Booking information: For general information and entry permits contact the Department of Wildlife and National Parks, P O Box 17, Kasane, Botswana (telephone Kasane 650235), or the Botswana Tourist Office, Private Bag 0047, Gaborone, Botswana (telephone Gaborone 353024). The Chobe Safari Lodge can be contacted at P O Box 10, Kasane (telephone Kasane 250336). It is not necessary to reserve public camping sites.

Etosha National Park

Location: Northern Namibia, 550 km north of Windhoek. Map page 256, reference: B9.

Wildlife: Lion, cheetah, elephant, giraffe, black rhino, black-faced impala, zebra, springbok, kudu, eland, gemsbok and more than 325 bird species.

Climate: Long, hot summers, with rainfall between December and April. Winter days mild, but nights cold.

When to go: All three camps are open all year round.

How to get there: There are two main routes to the park. A tarred road leads 550 km from Windhoek to the Von Lindequist Gate via Otjiwarongo and Tsumeb. The Namutoni rest camp lies 11 km from the gate. The second route leads 447 km from Windhoek, via Outjo, to the Andersson Gate and Okaukuejo rest camp. There are landing strips at all three camps and a private operator within the park organises transport for those visiting by air.

Opening time: Sunrise to sunset.

Getting around: A network of roads links the camps with more than 30 waterholes. Roads are gravel-surfaced and well-maintained. Petrol pump attendants at the rest camps can mend punctures, and there is a qualified mechanic at Okaukuejo for light, emergency repairs.

Special precautions: Take a course of anti-malaria tablets before visiting Etosha. As the distances in the park are so great, it is advisable to travel with water and emergency rations in case your vehicle breaks down. If it does, wait for help. It is not safe to leave your vehicle.

Accommodation: Etosha has three rest camps. Okaukuejo and Halali have luxury rondavels, bungalows with various facilities, 'bus quarters' for coach tourists, dormitories and tents. There is a caravan and camping site at Okaukuejo. Visitors to Namutoni either stay in the charmingly converted fort, built by the German authorities at the turn of the century to control cattle smuggling and the spread of rinderpest, or at the large camping and caravanning area. The private Mokuti Lodge, 500 m outside the Von Lindequist gate, offers accommodation ranging from luxury chalets to tourist units. It also has a runway for light aircraft.

Booking information: Enquiries and reservations should be made to Reservations, Private Bag 13267, Windhoek 9000, Namibia (telephone Windhoek 36975). For information and reservations at Mokuti Lodge, write to Central Reservations, Namib Sun Hotels, P O Box 2862, Windhoek 9000, Namibia (telephone Windhoek 33145).

Gonarezhou National Park

Location: On the Zimbabwe/Mozambique border, about 580 km southeast of Harare. Map page 256, reference: G10.

Wildlife: Elephant, buffalo, hippo, rhino, Lichtenstein's hartebeest, Burchell's zebra, giraffe, roan antelope and a rich birdlife.

Climate: Hot, wet summers, with temperatures often exceeding 40°C. Winters mild, although nights can be cold.

When to go: Open May to October (closed during the rainy season).

How to get there: Follow the Ngundu Halt road and turn right 20 km east of Chiredzi. The graded gravel road leads 59 km to Chipinda Pools on the northern side of the park. To reach the Mabalauta section and the Swimuwini Camp in the south, take the same road to Chiredzi and then follow the 160 km gravel road which runs along the western rim of the park. There is also a gravel road to Mabalauta from Nwanetsi (116 km).

Opening time: Gates open 6 am to 6 pm, but visitors must arrive not later than 5 pm.

Getting around: A network of game-viewing roads criss-crosses the park but, as their condition varies, a four-wheel drive vehicle is recommended. Check before you leave. Private air charters are available.

Special precautions: Take precautions against malaria and bilharzia. The elephants here are noted for their unpredictable behaviour – treat them with respect. There are no restaurants, shops or petrol stations in the park (the last point for fuel and stores is at Chiredzi), so be sure to take your own supplies. Although each chalet is equipped with a refrigerator, cooking utensils and linen, you must take your own crockery and cutlery.

Accommodation: Swimuwini Camp has five-bedded chalets and small three-bedded units with separate ablution facilities. There are small camping sites at Mabalauta, Chipinda Pools and Chinguli, as well as several undeveloped camping sites throughout the park. On the outskirts of the park, two safari lodges – Induna Lodge and Kwali Camp – offer more luxurious accommodation.

Booking information: National Parks Central Booking Office, Private Bag 8151, Causeway, Harare, Zimbabwe (telephone Harare 792731 or 792782/4); Cresta Hotels and Safaris, P O Box 2833, Harare (telephone Harare 794641); Zimbabwe Tourist Board, P O Box 9398, Johannesburg 2000 (telephone Johannesburg 3313137); Cane Air, P O Box 20, Chiredzi, Zimbabwe (telephone Chiredzi 643).

Greater St Lucia Wetland Park

(St Lucia Estuary, Charter's Creek, Fanie's Island, Mapelane, False Bay Park, Cape Vidal, Sodwana Bay National Park)

Location: Central Zululand coast, east and northeast of Mtubatuba and Hluhluwe. Map page 256, reference: G12.

Wildlife: Hippo, crocodile, black rhino, buffalo, kudu, waterbuck, reedbuck and more than 420 species of birds, including fish eagles, white pelicans, flamingoes and herons.

Climate: Subtropical with hot, humid summers and mild winters. Cold nights.

When to go: Throughout the year.

How to get there: St Lucia Estuary: Travel 26 km on the R620 from Mtubatuba to the estuary. To reach Charter's Creek and Fanie's Island take the N2 from Mtubatuba for 25 km and turn right at the Nyalazi River Halt turnoff. To get to Mapelane, turn right on the N2, 30 km north of Empangeni. False Bay Park: Take the N2 north of Mtubatuba for 55 km, turn right to Hluhluwe Village and follow the signposts to the reserve. Cape Vidal: Continue northwards from the St Lucia Estuary into the Mfabeni section (previously the Eastern Shores State Forest); Cape Vidal reserve is 30 km further north. Sodwana Bay National Park: Turn right off the N2 at Mhlosinga and follow the signposts for 80 km.

Opening time: Varies within the reserve. It is best to check before leaving.

Getting around: There are a number of trails and fascinating drives throughout the Greater St Lucia Wetland Park, as well as boat trips and launch tours in the estuary. For details contact the Natal Parks Board.

Special precautions: Take anti-malaria precautions before entering the area. Beware of crocodiles and hippos.

Accommodation: Charter's Creek, Fanie's Island and Mapelane: cabins and rest huts; there are camping facilities at Fanie's Island and Mapelane. False Bay Park: four-bedded rustic huts on Dugandlovu Trail; camping and caravanning facilities. Cape Vidal: log cabins, a bush camp, anglers' cabins and a camping and caravan site. Sodwana Bay National Park: fully equipped five- and eight-bedded chalets, and open camping sites.

Booking information: The Reservations Officer, Natal Parks Board, P O Box 662, Pietermaritzburg 3200 (telephone Pietermaritzburg 471981).

Hluhluwe Game Reserve

Location: KwaZulu, Natal. 280 km north of Durban. Map page 256, reference: G12.

Wildlife: Black and white rhino, elephant, buffalo, zebra, leopard, cheetah, hippo, crocodile, kudu, impala, reedbuck, blue wildebeest, duiker and giraffe. Over 350 bird species include bateleur eagle, Narina trogon, white-backed vulture and ground hornbill.

Climate: Hot, wet summers and balmy winters.

When to go: Throughout the year.

How to get there: Follow the N2 from Durban to Hluhluwe Village and turn left onto the tarred road to the Memorial Gate. A second route from the N2 lies 3,4 km north of the Mtubatuba junction. Turn left for Nongoma/Hluhluwe Game Reserve; 21 km along this road, turn north onto a gravel road which leads to the Gunjaneni Gate.

Opening time: Sunrise to sunset. Telephone beforehand as these times change seasonally.

Getting around: There are game-viewing drives within the reserve, with two self-guided auto trails. Guided walks are also available. Visitors may leave their cars only at designated viewpoints and in the camps.

Special precautions: Anti-malaria tablets should be taken before visiting the park. Bathing in the rivers or dams is prohibited throughout the year because of the presence of bilharzia and crocodiles.

Accommodation: Hilltop Camp has three 6-bedded cottages and 20 two-bedded rondavels. Mtwazi Lodge offers more luxurious accommodation, and for those who want unspoilt privacy, there is the rustic bush camp of Muntulu. There are picnic and braai facilities at Hilltop Camp and Gunjaneni Gate for day visitors.

Booking information: Address enquiries and reservations to Reservations Officer, Central Booking Office, Natal Parks Board, P O Box 662, Pietermaritzburg 3200 (telephone Pietermaritzburg 471981).

Hwange National Park

Location: Western Zimbabwe, about 280 km northwest of Bulawayo. Map page 256, reference: E/F9.

Wildlife: Elephant, buffalo, rhino, lion, leopard, cheetah, giraffe, sable, kudu, impala, steenbok, zebra and a wide variety of bird species.

Climate: Hot summers, with rain between October and March. Dry, sunny winters.

When to go: With the exception of Robins Camp, which is closed during the rainy season (November to April), the park is open throughout the year. However, the peak game-viewing period is September and October.

How to get there: Follow the main Bulawayo-Victoria Falls road for approximately 264 km to the turnoff to Main Camp. A 24 km tarred road leads to the camp. The turnoff to Sinamatella Camp lies a further 86 km along the Bulawayo-Victoria Falls road, with a 45 km drive to the camp. Air Zimbabwe offers flights to Hwange National Park Aerodrome near Main Camp, and private/charter aircraft may use the strip at Main Camp itself (permission is required). Visitors by rail may travel to Dete station, where a coach service to Main Camp is offered.

Opening time: Sunrise to sunset.

Getting around: Most of the camps offer game-viewing drives – some by moonlight – along the 480 km network of roads within the park. Visitors may also take escorted foot safaris.

Special precaution: Anti-malaria tablets should be taken before visiting Hwange.

Accommodation: There are three National Parks tourist camps – Main, Robins/Nantwich and Sinamatella – with serviced cottages, lodges, chalets and camping and caravanning facilities. Bumbusi and Lukosi Exclusive camps offer four 2-bedded units and a small cottage. Deka Exclusive Camp has serviced family units. Hwange Safari Lodge, just outside the park near Main Camp, offers luxury accommodation and tours as does Ivory Lodge, a similar luxury establishment. Small groups may, with the permission of the warden, camp overnight at some of the enclosed picnic sites within the reserve.

Booking information: National Parks Board Central Booking Office, Private Bag 8151, Causeway, Harare, Zimbabwe (telephone Harare 792731 or 792782/4). Private lodges and camps can be reserved by contacting United Touring Company, P O Box 2914, Harare (telephone Harare 793701).

Itala Game Reserve

Location: Northern Natal, 70 km northeast of Vryheid. Map page 256, reference: G12.

Wildlife: Elephant, white and black rhino, blue wildebeest, red hartebeest, giraffe, leopard, cheetah, tsessebe, zebra, impala, kudu, eland, klipspringer and warthog. Raptors include martial eagle, black eagle, Wahlberg's eagle, fish eagle, black-breasted snake eagle and the rare southern banded snake eagle.

Climate: Winters mild, with cold nights. Summers warm to hot.

When to go: Throughout the year.

How to get there: Take the R69 from Vryheid to Louwsburg, and follow the signs to Itala.

Opening time: Sunrise to sunset.

Getting around: Five-day wilderness trail (March to October). Day walks. There are 30 km of gravel game-viewing roads, including the Ngubhu Loop Road, and various picnic spots.

Special precaution: A course of anti-malaria tablets is recommended before visiting Itala.

Accommodation: Ntshondwe Camp: Two 5-bedded chalets, one luxury lodge with three en suite bedrooms. There are three bush camps: Thalu, a four-bedded camp, Mbizo, an eight-bedded camp and Mhlangeni, which sleeps 10 (take your own food, drink, torches and towels).

Booking information: The Reservations Officer, Natal Parks Board, P O Box 662, Pietermaritzburg 3200 (telephone Pietermaritzburg 471981). For camping, contact the Officer-in-Charge, Itala Nature Reserve, P O Box 42, Louwsburg 3150 (telephone Louwsburg 1413).

Kafue National Park

Location: Western Zambia, about 330 km west of Lusaka. Map page 256, reference: E7/8.

Wildlife: Lion, leopard, elephant, hippo, buffalo, sable, roan antelope, wild dog, spotted hyaena and a rich birdlife.

Climate: Hot, wet summers (November to April) and dry, mild winters.

When to go: Some camps remain open all year, while others close between October or November and May. Check before visiting the park.

How to get there: Follow the main road leading west from Lusaka for 147 km to Mumbwa. Continue towards Kaoma for a further 66 km until you reach the turnoff to Lake Itezhi-Tezhi. Follow this road for 115 km to reach the park. Should you come from Livingstone in the south, follow the main road north to Lusaka for 126 km until you reach Kalomo. Turn left at Kalomo for the Ndumdumwense Gate, some 80 km further on.

Opening time: Sunrise to sunset.

Getting around: There is a network of game-viewing roads. Private operators offer game-viewing, bird-watching and fishing trips. You may need a four-wheel drive vehicle during the rainy season.

Special precautions: Take the normal precautions against malaria and bilharzia. Check with your doctor about medical and vaccination requirements for cholera and yellow fever.

Accommodation: Ranges from self-catering to luxury camps. Busanga Trails runs the luxurious Lufupa, Moshi, Safari and Ntemwa camps, as well as four camps in the northern section of the park. Musungwa Safari Lodge is privately run. At Lake Itezhi-Tezhi, Zambia's Wildlife Conservation Society owns a self-catering camp which, although private, will admit visitors who wish to take out temporary membership. Lubungu Wildlife Safaris runs Hippo Camp in the north.

Booking information: The Zambia National Tourist Board, P O Box 30017, Lusaka, Zambia (telephone Lusaka 229087) or P O Box 591232, Kengray, Johannesburg 2100 (telephone Johannesburg 6229206/7). Busanga Trails Ltd, P O Box 37538, Lusaka (telephone Lusaka 222075). The Wildlife Conservation Society of Zambia, P O Box 30255, Lusaka (telephone Lusaka 254226). Lubungu Wildlife Safaris, P O Box 31701, Lusaka (telephone Lusaka 251823).

Kalahari Gemsbok National Park

Location: Northwestern corner of South Africa, between Namibia and Botswana. Map page 256, reference: D11.

Wildlife: Lion, leopard, cheetah, gemsbok, red hartebeest, blue wildebeest, steenbok, springbok and more than 200 species of birds.

Climate: Summers hot and dry. Winters warm, but nights cold, often below freezing point.

When to go: All year round but the best time for viewing game is from February to May.

How to get there: The most popular approach route is a 288 km gravel road cutting north from the tarred highway between Upington and Namibia. The turnoff is 63 km from Upington, and petrol is available at Noenieput and Andriesvale. A second route approaches Andriesvale from the east by way of Kuruman, Hotazel and Vanzylsrus. There are landing strips at the Twee Rivieren, Mata Mata and Nossob rest camps (contact the park authorities for permission to use them).

Opening time: The park's three rest camps open their gates at sunrise and close at sunset (with an hour's break at midday), seven days a week.

Getting around: Seen from the north, the park's road system has the shape of a giant 'A', consisting of the two roads along the riverbeds and a dune road connecting unfenced 'halfway house' picnic spots. The roads are in good condition but you may not leave them because you may get stuck in soft sand. Before you set out on a journey, state your route and destination. If you fail to arrive by sunset, the park authorities will launch a search. Petrol is available at all three camps.

Special precautions: Take a course of anti-malaria pills before entering the park. When going for a drive, take water and emergency rations in case you have a breakdown.

Accommodation: There are rest camps at Mata Mata in the west, Twee Rivieren in the south, and Nossob towards the north. All three camps offer family cottages (two double bedrooms, kitchen and bathroom); self-contained huts (up to four beds, kitchen and bathroom); and ordinary huts (up to four beds, with access to a shared kitchen and shared bathroom). Bedding, towels, cooking and eating utensils are supplied. Each camp has a camping and caravan site with braai and washing-up facilities, and an ablution block.

Booking information: National Parks Board, P O Box 787, Pretoria 0001 (telephone Pretoria 3431991); or the National Parks Board, P O Box 7400, Roggebaai 8012 (telephone Cape Town 222810).

Kruger National Park

Location: On the eastern Transvaal border with Mozambique, extending from Zimbabwe to Swaziland. Map page 256, reference: G10/11.

Wildlife: Elephant, roan antelope, tsessebe and eland occur mainly in the north; lion, giraffe, impala, Burchell's zebra and black rhino prefer the south. Other species in the park include hippo, crocodile, leopard, cheetah, wild dog, buffalo, blue wildebeest, eland, kudu, sable, a variety of smaller antelope and carnivores and more than 400 species of birds.

Climate: Hot summers with rain; cool, dry winters. The highest rainfall occurs in the south and southwest.

When to go: Many people prefer winter with its mild days and cool nights. More animals tend to concentrate around the waterholes in the dry season.

How to get there: There are eight public entrance gates, of which Malelane, Phalaborwa and Crocodile Bridge are popular choices. To get to Malelane, take the N4 from Nelspruit. After 70 km take the signposted left turn to the gate, which is 3 km further on. Berg-en-dal, 12 km away, is the nearest large camp. To get to the Phalaborwa entrance, travel eastward on the R71 from Pietersburg for 214 km. On the way you will pass through Tzaneen and Phalaborwa. The nearest camp, Letaba, is 50 km further on. To get to Crocodile Bridge, take the N4 from Nelspruit towards Komatipoort for about 110 km before turning left. The gate is about 10 km from the turnoff.

Opening time: January to February 5.30 am to 6.30 pm; March 5.30 am to 6 pm; April 6 am to 5.30 pm; May to August 6.30 am to 5.30 pm; September 6 am to 6 pm; October 5.30 am to 6 pm; November to December 5.30 am to 6.30 pm.

Getting around: A good, reliable network of roads puts visitors in touch with all sections of the park, which is 350 km long and covers 19 455 sq. km. Main roads are tarred and secondary roads are gravelled. There is a general speed-limit of 40-50 km/h. Outside the camps, visitors must stay in their cars, except at designated picnic spots or monuments. There are filling stations at all the gates, except Malelane, Jock-of-the-Bushveld and Paul Kruger, and at the major rest camps. Seven wilderness trails – the Wolhuter, Bushman, Olifants, Nyalaland, Metsi-Metsi, Sweni and Napi – enable visitors to see the bush on foot and to camp out overnight at bush camps where food, bedding and cooking and eating utensils are provided. These trails are open to a maximum of eight people at a time and are led by an experienced game ranger.

Special precautions: Take a course of anti-malaria tablets before entering the park. In summer, use a sunscreen to protect your skin from the sun's ultra-violet rays.

Accommodation: There are 24 rest camps at Kruger, which range from the largest and most sophisticated at Skukuza, to one for campers only. Most of the camps offer a variety of accommodation, from six-bedded family units to two- and three-bedded huts equipped with bathroom, refrigerator and air-conditioning. Skukuza has four- and six-bedded family cottages, two- and three-bedded huts, seven self-contained guest cottages, dormitory accommodation for school groups and a caravan and camping site with full ablution facilities. There are also private, fully equipped camps, such as Boulders north of Letaba, that must be booked as a unit by one group. Supplies can be obtained at Phalaborwa or at Letaba. The larger camps include Olifants, Satara and Letaba, while among the smaller camps are Punda Maria, Crocodile Bridge, Balule and Nwanetsi.

Booking information: National Parks Board, P O Box 787, Pretoria 0001 (telephone Pretoria 3431991); or the National Parks Board, P O Box 7400, Roggebaai 8012 (telephone Cape Town 222810).

Makgadikgadi Pans Game Reserve and Nxai Pan National Park

Location: Northern Botswana, about 350 km west of Francistown. Map page 256, reference: E9.

Wildlife: Lion, cheetah, leopard, wild dog, wildebeest and zebra. Vast numbers of waterbirds, particularly flamingoes, when good rains fall.

Climate: Hot, wet summers (rain between January and April); cool, dry winters, with clear days and cold nights.

When to go: June to late September, when the nights are cool, are the best months for the Makgadikgadi Pans, but the grasslands of Nxai Pan teem with game only in the rainy months (January to April). The reserves are open all year.

How to get there: There are scheduled air services from Harare and Johannesburg to Gaborone, which in turn is linked to Francistown and Maun. The 350 km northern road to the reserves from Francistown is suitable for saloon cars and is tarred for its entire length. However, four-wheel drive vehicles are essential to enter the reserves.

Opening time: 6 am to 6 pm.

Getting around: The roads in the pans are at best tracks. Essential equipment includes puncture repair kit, extra fuel in metal (not plastic) jerry cans, at least 20 litres of water, motor oil, jacks, spanners, fanbelts, inner tubes, a pump, spare coil, rotor, fuel-pump kit, regulator, condenser, plugs, points, radiator hose, nylon tubing, clips and strips of steel mesh wire (for use when bogged down in sand). A guide should accompany you in the parts of the Makgadikgadi outside the reserve. A compass could be useful.

Special precautions: Start a course of anti-malaria tablets and take sufficient food and water for the duration of your trip. Bilharzia is a constant threat, so do not swim.

Accommodation: There is no organised accommodation in the pans, but there are two public camping sites in the Nxai Pan park, and two at Makgadikgadi Pans.

Booking information: You don't have to book camping sites in advance. For information, write to the Department of Wildlife and National Parks, P O Box 131, Gaborone, Botswana (telephone Gaborone 371405), or the Botswana Tourist Office, Private Bag 0047, Gaborone (telephone Gaborone 353024).

Mana Pools National Park

Location: Northeastern Zimbabwe, some 300 km north of Harare. Map page 256, reference: G8

Wildlife: Elephant, hippo, buffalo, rhino, nyala, zebra, kudu, impala, eland, lion, leopard, hyaena, wild dog, crocodiles and numerous bird species.

Climate: Subtropical, with hot, rainy summers and mild winters.

When to go: Open May to October. Check these dates in advance – they may change with weather conditions. Note also that the number of visitors is strictly controlled – no more than 50 vehicles are allowed into the park at any one time.

How to get there: Follow the Harare-Chirundu road (tarred) northwest for 312 km to Marongora (Mana Pools permits obtainable here); continue for another 6 km and turn right onto a gravel road leading 30 km to Nyakasikana Gate. Nyamepi Camp lies 42 km beyond the gate. There is a bush airstrip at Marongora. Note that the last petrol stop is at Makuti, some 16 km south of Marongora.

Opening time: Sunrise to sunset.

Getting around: Roads within the park are often in poor condition. Visitors are allowed to leave their vehicles at their own risk, but the authorities stress that caution must be exercised, and feeding the animals is forbidden. There are game-viewing drives, canoes can be hired, or you can take one of the land or river photographic safaris offered at the luxury camps on the western border of the park. Travel within the park is only allowed between half an hour before sunrise and half an hour before sunset.

Special precautions: Take precautions against malaria and bilharzia. There are tsetse flies here, and all drinking water should be boiled or treated chemically. Beware of dangerous animals.

Accommodation: Nyamepi Camp has a caravan and camping site with ablution facilities and, nearby, two fully equipped eight-bedded lodges; 13 km from Nyamepi is Vundu Camp, with huts accommodating 12 people, and communal lounge, kitchen and ablution facilities. The more remote Nkupe and Mucheni camps are camping sites with ablution facilities and fireplaces, each available to a single party of up to 12 people with no more than two vehicles. On the western boundary, the two more luxurious safari camps, Chikwenya and Ruckomechi, both offer similar facilities and safaris.

Booking information: For accommodation, contact the National Parks Board Central Booking Office, Private Bag 8151, Causeway, Harare, Zimbabwe (telephone Harare 792731 or 792782/4). Fothergill Island Safaris, Private Bag 2081, Kariba, Zimbabwe (telephone Kariba 2253 or Harare 705144). To reserve accommodation at Chikwenya and Ruckomechi, contact the United Touring Company, P O Box 2914, Harare (telephone Harare 793701).

Masai Mara National Reserve

Location: Southwestern Kenya, 275 km west of Nairobi. Map page 256, reference: H3.

Wildlife: Lion, cheetah, leopard, elephant, buffalo, rhino, hippo, crocodile, wildebeest, topi, eland, zebra, giraffe and a rich birdlife. Between July and September, hundreds of thousands of wildebeest migrate northwards into the reserve from the Serengeti plains. They graze here until the rains break in October or November, when they return southwards.

Climate: Summers hot, with rains in November and December (the 'short' rains), and from March to May (the 'long' rains).

When to go: Open all year.

How to get there: Follow the main road from Nairobi to Narok (last place to get petrol before the reserve). At about 20 km after Narok, take either the tarred C12 road on the left which leads to the eastern section of the reserve, or turn right onto a rugged road which leads to the western part of the reserve. This road is bad during the rainy season, and a four-wheel drive vehicle is necessary. There is also a twice-daily air service from Nairobi. Each of the major luxury lodges has an airstrip.

Opening time: 6 am to 6 pm.

Getting around: The reserve is crisscrossed by a network of tracks, but visitors are allowed to drive off the roads. During the rainy season, you will need a four-wheel drive vehicle if you wish to stray from the main graded tracks. Game-viewing safaris are available at the various camps, and hot-air balloon trips provide a unique view of the game.

Special precautions: Take precautions against malaria and bilharzia. Use tap water only for ablutions. Most lodges and camps offer filtered water, although bottled water can be bought. Purify all other water by either boiling it or using water-purifying tablets.

Accommodation: There is a variety of luxury lodges, luxury tented camps and less expensive tented and banda (hutted) camps within the reserve, owned mostly by local hotel chains. Some provide accommodation for tour groups, while others are more remote. It is essential to make reservations well in advance.

Booking information: Abercrombie and Kent, P O Box 59749, Nairobi, Kenya (telephone Nairobi 334955). There is also an information bureau outside the Hilton Hotel, P O Box 30624, Nairobi (telephone Nairobi 334000). Wildlife Conservation and Management Department, P O Box 40241, Langata Road, Nairobi (telephone 501081). Balloon Safaris Ltd, P O Box 47557, Nairobi (telephone Nairobi 335807). Pollman's Tours and Safaris, P O Box 84198, Mombasa, Kenya (telephone Mombasa 312565/6/7).

Mkuzi Game Reserve

Location: About 335 km northeast of Durban. Map page 256, reference: G12.

Wildlife: Black rhino, white rhino, giraffe, nyala, blue wildebeest, warthog, kudu and smaller antelope. More than 400 species of birds.

Climate: Hot, humid summers; mild winters with cold nights.

When to go: The reserve is open daily throughout the year.

How to get there: Visitors arriving from the south should take the signposted turnoff to the reserve along the main north coast road (about 35 km north of Hluhluwe Village). Follow the gravel road for 15 km and take the signposted turnoff through the Lebombo mountains: the hutted camp is 15 km from this point. Visitors from the north should travel to Mkuze Village, which is 18 km from the reserve's entrance gate and 28 km from Mantuma Camp. The road is well signposted.

Opening time: From 8 am to 12.30 pm, and 2 pm to 4.30 pm. Sundays: 8 am to 12.30 pm, and 2 pm to 4 pm.

Getting around: An 80 km network of roads leads through a variety of bushveld country, offering excellent game-viewing. Day walks are conducted by a game guard (make arrangements a day in advance with the camp superintendent).

Special precaution: Start a course of malaria tablets before visiting the reserve.

Accommodation: The Mantuma hutted camp comprises six 3-bedded rest huts with communal kitchen and ablution blocks, five 5-bedded bungalows, four 3-bedded bungalows and two self-contained cottages, each with seven beds. Visitors may also stay in two 7-bedded cottages. There are four rustic huts with an ablution block (hot and cold water), and a camping and caravan site is situated at the entrance gate. Two bush camps, Nhlonhlela (hutted) and Umkumbi (tented) offer peace and seclusion.

Booking information: You can book camping and caravan sites through the Camp Superintendent, Mkuzi Game Reserve, P O Mkuze 3965 (telephone Mkuzi Reserve Call Office). Address reservations for the rustic and main camps to the Reservations Officer, Natal Parks Board, P O Box 662, Pietermaritzburg 3200 (telephone Pietermaritzburg 471981).

Moremi Wildlife Reserve

Location: Northwestern Botswana, 98 km northwest of Maun. Map page 256, reference: D/E9.

Wildlife: Elephant, hippo, lion, leopard, wild dog, baboon, serval, buffalo, warthog, tsessebe, kudu, impala, roan, reedbuck, lechwe, waterbuck and about 300 species of birds.

Climate: Summers hot and humid with rain; winters warm with cold nights.

When to go: Throughout the year, but heavy rains in summer can make travelling conditions very difficult.

How to get there: From Maun, follow track northeastwards toward Chobe National Park (road is tarred from Maun to Shorobe Village). After 64 km turn left and drive another 34 km to the Moremi south gate. From Savuti, drive southwards and turn right at the signpost to the reserve. The Savuti road is very sandy, and obstacles in the road, such as fallen trees, are common. Four-wheel drive vehicles are necessary for getting to Moremi by road. Light aircraft from Maun are a popular way of getting there.

Opening time: 6 am to 6 pm.

Getting around: You will need to travel in a four-wheel drive vehicle for game-viewing drives in Moremi. Boat trips can be made from Xakanaxa Camp near Third Bridge.

Special precautions: Take precautions against malaria and bilharzia. Check condition of roads. No swimming is allowed in park's waters. Watch out for crocodiles and other dangerous animals.

Accommodation: Several safari companies operate in and around the Moremi Wildlife Reserve, and advance booking is essential. Most travel agents will supply the names and addresses of these companies. There are public camping sites at South Gate, North Gate and Third Bridge and two chalet camps at San-ta-wani and Khwai River. Although there is available water, it is probably safer to take your own if you can.

Booking information: The Department of Wildlife and National Parks, P O Box 111, Maun, Botswana (telephone Maun 260230). For more information about the park, contact the Botswana Tourist Office, Private Bag 0047, Gaborone, Botswana (telephone Gaborone 353024). For bookings at San-ta-wani and Khwai River contact Safariplan, P O Box 4245, Randburg 2125 (telephone Johannesburg 8861810).

Namib-Naukluft Park

Location: Namib Desert coastal region, between Swakopmund and Lüderitz. Map page 256, reference: B10/11.

Wildlife: Gemsbok, springbok, ostrich and mountain zebra.

Climate: Ranges between temperate coastal to desert, with hot summers and moderate winters.

When to go: Open all year round, but beware of high temperatures between October and March.

How to get there: From Windhoek, the most popular approach is via the main road to Swakopmund, but a more scenic route crosses the Gamsberg Pass before continuing through the park to Walvis Bay. To reach Naukluft and the southern dunes, drive by way of Solitaire (249 km from Windhoek), then south on the road to Sesriem.

Opening time: Sunrise to sunset.

Getting around: If you stick to the proclaimed main roads, you may drive through the park without a permit. Petrol is available at Sesriem and Solitaire; but otherwise motorists should fill up at Walvis Bay or Swakopmund. A brochure gives directions for the interesting four-hour Welwitschia drive (you need a permit, available from the Directorate of Nature Conservation and Tourism in Swakopmund). The 120 km Namib-Naukluft Hiking Trail takes eight days, although a shorter four-day route is also available. Hiking, in groups of three to 12, is permitted between March and October. Only water, toilet facilities and basic shelter are provided. There are shorter day trails.

Special precautions: Take all the food and water (extremely important) you will need, and emergency supplies in case your vehicle breaks down. Although firewood is available at Sesriem and Naukluft, you should also pack your own if you are headed for one of the smaller sites.

Accommodation: There is no permanent accommodation in the park but there are a number of official overnight camping sites equipped with picnic tables and toilets. The sites include Homeb (in the bed of the Kuiseb River, 20 km up river from Gobabeb); Ganab (near a watering point that is a good place to watch game); Welwitschiavlakte (best known for the fascinating *Welwitschia mirabilis*); Naukluft (available only to those with reservations); and Sesriem.

Booking information: Reservations, Private Bag 13267, Windhoek 9000, Namibia (telephone Windhoek 36975). You may also apply to the Head: Tourism, Municipality, Private Bag 5017 Swakopmund 9000, Namibia (telephone Swakopmund 2807/8).

Ndumo Game Reserve

Location: 470 km from Durban in the extreme northeast of Natal. Map page 256, reference: G12.

Wildlife: There are over 400 bird species in Ndumo, almost as many as in the Kruger National Park, which is 190 times larger. The game includes hippo, crocodile, impala, nyala, bushbuck, grey duiker, white rhino, black rhino, buffalo, zebra, red duiker, giraffe, reedbuck, red musk shrew, fruit bat, pangolin (scaly anteater), forest dormouse, vlei rat, striped polecat, water mongoose, aardwolf and aardvark.

Climate: Moderately dry, subtropical climate, with 30 per cent less rain than at the coast.

When to go: Throughout the year. The hottest month is February, the coolest July. The bird population is greatest during summer.

How to get there: Turn off the N2 at Nkonkoni railway siding, 8 km north of Mkuze Village, and travel 21 km to the Jozini Dam and village. The road to Ndumo, which crosses the dam wall, is clearly signposted. It is tarred until the last 20 km. There is petrol at a store near the Ndumo reserve, Jozini and Ingwavuma.

Opening time: All visitors must report to the camp superintendent's office on arrival (before 4.30 pm). The entrance gate is locked from sunset to sunrise.

Getting around: Depending on demand, morning and afternoon tours are conducted around the pans (make arrangements with the camp superintendent). Other areas are open to normal motoring, and a self-guided auto trail brochure is available. You can take a day walk, accompanied by a game guard.

Special precaution: Anti-malaria precautions are advisable.

Accommodation: There is no caravan or camping site. There are seven 3-bedded huts, or squaredavels, each with a gas-powered refrigerator, and a communal kitchen and chef. Bedding, cutlery and crockery are provided. There is electricity from 5 pm to 10.30 pm.

Booking information: The Central Reservations Officer, KwaZulu Bureau of Natural Resources, 367 Loop Street, Pietermaritzburg 3201 (telephone Pietermaritzburg 946698).

Pilanesberg National Park

Location: Southwestern Transvaal, about 55 km north of Rustenburg, and 187 km northwest of Johannesburg. Map page 256, reference: F11.

Wildlife: Elephant, black and white rhino, buffalo, giraffe, leopard, cheetah, antelope, zebra, hippo and more than 320 bird species, including the black eagle and the fish eagle.

Climate: Hot summers, with rain; cool winters.

When to go: Throughout the year.

How to get there: Take the R510 north of Rustenburg in the direction of Thabazimbi for 56 km, then turn left at the signpost to the reserve. Manyane Gate is 6 km from the turnoff.

Opening time: From April to August 5.30 am to 6 pm; from September to March 5 am to 6 pm.

Getting around: There is a 100 km network of roads. Organised game-viewing drives and wilderness trails are available. Game-viewing hides dot the park, with the Kwa Maritane hide accessible via an underground tunnel.

Accommodation: There is a hotel with beautiful chalets overlooking the park at Kwa Maritane. Tented accommodation is provided at Manyane, Mankwe, Kololo and Metswedi camps. Manyane caravan and camping site has 100 stands and ablution facilities. At Bosele Camp, wooden huts accommodate groups of up to 180. There are restaurants at Manyane and Pilanesberg Centre, where food, books and curios may also be purchased.

Booking information: The Reservations Officer, Pilanesberg National Park, P O Box 1201, Mogwase 0302, Bophuthatswana (telephone 01460-55351 or 22377/8).

Royal Natal National Park

Location: On the northern crest of the Natal Drakensberg, about 390 km northwest of Durban. Map page 256, reference: F12.

Wildlife: There are over 180 species of birds, including the bearded vulture, Cape vulture and black eagle. Mammals include hare, klipspringer, baboon, bushbuck, mountain reedbuck, black-backed jackal, porcupine, grey rhebok, eland and grey duiker.

Climate: Warm to hot summers; cold winters, with frost in June and July. Frequent afternoon thunderstorms in summer (October to March).

When to go: Throughout the year. Try to avoid the Christmas and Easter holidays when the park is very popular.

How to get there: The park is 45 km (tarred) west of Bergville, or 72 km from Harrismith via the Sterkfontein Dam and Oliviershoek Pass.

Opening time: The Tendele Camp and Royal Natal National Park campsite gates are open between sunrise and sunset.

Getting around: Thirty-one walks and climbs have been mapped out, ranging from 3 km to 45 km. You can also explore the region on horseback: there are several bridle path trails, and horses to hire.

Special precautions: If you want to climb to the top of the berg, or try any difficult climbs, obtain permission from the reserve warden. Rescues are organised from the warden's office (fatal accidents still occur). Don't walk or climb alone, and if mist comes up, stay put until it clears.

Accommodation: The Tendele hutted camp can accommodate up to 60 people in a variety of bungalows, chalets and cottages. Tendele Lodge provides luxury accommodation for six people in three en-suite bedrooms. Apart from the Tendele bungalows and cottages and the camping sites at Mahai, accommodation includes the Royal Natal National Park Hotel, P O Mont-aux-Sources 3353 (telephone Bergville 381051).

Booking information: The Reservations Officer, Natal Parks Board, P O Box 662, Pietermaritzburg 3200 (telephone Pietermaritzburg 471981). The camping sites at Mahai are booked through the Officer-in-Charge, Royal Natal National Park, P O Mont-aux-Sources 3353 (telephone Jagersrust 4386423).

Sabi Sand Game Reserve

Location: The southwestern boundary of Kruger National Park, about 500 km northeast of Johannesburg. Map page 256, reference: G11.

Wildlife: Elephant, black and white rhino, hippo, crocodile, buffalo, lion, cheetah, leopard and a variety of antelope.

Climate: Hot summers, cold winters.

When to go: Throughout the year.

How to get there: Take the R40 north of Nelspruit, pass through Hazyview and take the R536 (Skukuza Highway) towards the Paul Kruger Gate. After about 36 km a left turn will lead you to the signposts for Sabi Sabi, Londolozi, Mala Mala, Kirkman's Kamp, Harry's Camp, Trekker Trails and Inyati.

Opening time: Sunrise to sunset.

Getting around: Game-viewing drives in open vehicles, including night drives, and accompanied walking trails. Fishing trips are available at Inyati.

Special precaution: Take a course of anti-malaria pills before entering the park. Mala Mala offers malaria tablets on arrival.

Accommodation: Mala Mala is the flagship of the private game reserves in Sabi Sand. Within it are the luxury Mala Mala main camp; Kirkman's Kamp (with one- and two-bedroomed cottages); Harry's Camp (with medium-priced accommodation); and Trekker Trails – a bush camp, accommodating six guests at a time. Sabi Sabi: the two lodges have a total of 45 two-bedded chalets. Londolozi: the main camp has luxury thatched chalets and rondavels; two other camps have luxury bedrooms and cabins. Inyati: Inyati Game Lodge has nine 2-bedded luxury chalets.

Booking information: Mala Mala, Kirkman's Kamp, Harry's Camp and Trekker Trails: Rattray Reserves Reservations, P O Box 2575, Randburg 2125 (telephone Johannesburg 7892677/8/9). Sabi Sabi: P O Box 52665, Saxonwold 2132 (telephone Johannesburg 8804840). Londolozi: P O Box 1211, Sunninghill Park 2157 (telephone Johannesburg 8038421). Inyati: P O Box 38838, Booysens 2016 (telephone Johannesburg 4930755).

Serengeti National Park

Location: Northeast Tanzania, about 200 km east of Arusha. Map page 256, reference: H3.

Wildlife: Lion, cheetah, leopard, wild dog, hyaena, caracal, golden and black-backed jackal, wildebeest, zebra, gazelle and a wealth of birdlife.

Climate: Hot summers, mild winters. 'Long' rains between April and May, and 'short' rains in November and December.

When to go: Open all year, with the most popular season being between December and May, when vast herds of game swarm to the southern parts of the park.

How to get there: Follow the main road westwards from Arusha (the roads are in poor condition). Accessible by air from Dar es Salaam, Arusha and Nairobi and by road from Arusha. Tours offered by various private operators.

Opening time: Sunrise to sunset.

Getting around: Rough game-viewing tracks crisscross the park, but some are suitable for four-wheel drive vehicles only.

Special precautions: Take the usual precautions against malaria. Bilharzia is present in rivers and lakes. Drink only water which has been boiled or purified with tablets.

Accommodation: At Seronera, the park headquarters, a hostel offers dormitory-style accommodation; the luxury Seronera Lodge (1 km from the village) caters for the well-heeled and, nearby, there is a campsite. Another lodge and camping facilities are at Ndutu on the Serengeti/Ngorongoro Crater border. A number of other camping sites are situated throughout the park.

Booking information: Tanzanian National Parks, P O Box 3134, Arusha, Tanzania (telephone Arusha 3471). Subzali Tours and Safaris, P O Box 3061, Arusha (telephone Arusha 3681). Arumeru, P O Box 730, Arusha (telephone Arusha 7637). Wilderness Safaris, P O Box 651171, Benmore, South Africa 2010 (telephone Johannesburg 8841458). Holiday Tours, P O Box 4942, Randburg, South Africa 2125 (telephone Johannesburg 7870512).

Skeleton Coast National Park

Location: Northern Namib Desert coastal strip, 205 km north of Swakopmund. Map page 256, reference: A9.

Wildlife: Springbok and gemsbok are common at the mouths of the rivers, jackal and brown hyaena scavenge on the beach and many waterfowl and other birds can be seen on vleis in the deltas. In the northern wilderness area, elephant and black rhino come within 10 km of the coast, and lion and cheetah have been seen on the sand flats.

Climate: Summers dry and hot, winters generally cool, with morning and evening fog. Easterly winds can bring sudden high temperatures.

When to go: Throughout the year. The most popular months are December and January.

How to get there: From Swakopmund, drive north up the coast road to the park entrance gate at Ugab. Torra Bay is 114 km further north and Terrace Bay is 48 km beyond that. From the east, approach the park from Khorixas and enter at Springbokwater gate, then continue to Torra Bay (238 km from Khorixas) and on to Terrace Bay. There is an airstrip at Terrace Bay.

Opening time: Both gates open at sunrise, and departing visitors may pass through them at any time before sunset. Arriving visitors must reach Ugab or Springbokwater gates not later than 3 pm to allow enough time to drive to Terrace Bay. To gain access to the park, you will be asked to produce receipts showing that you have booked accommodation. Alternatively, a day permit for passing through between Ugab and Springbokwater may be taken out at any of the gates.

Getting around: Although the gravel roads within the park are suitable for saloon cars, you are advised to carry fuel, water and basic spares. Fuel and a tyre-repairing service are available at Terrace Bay.

Special precautions: Take with you a comprehensive map of the area. Dunes sometimes shift across the road, so don't drive too fast.

Accommodation: Terrace Bay's visitors are accommodated in rooms of the old mining camp (two or three beds and shower). The accommodation tariff provides for full board, with meals served in the licensed restaurant. Although there is a fully equipped shop, there are no camping facilities at Terrace Bay. The large caravan and camping site at Torra Bay is open during December and January.

Booking information: The Director of Nature Conservation, Reservations, Private Bag 13267, Windhoek 9000, Namibia (telephone Windhoek 36975). To book a fly-in safari, write to Skeleton Coast Safaris, P O Box 2195, Windhoek 9000 (telephone Windhoek 224248).

Timbavati Private Nature Reserve

Location: Eastern Transvaal, 500 km northeast of Johannesburg; 200 km northeast of Lydenburg. Map page 256, reference: G11.

Wildlife: White lions, leopard, elephant, white rhino, giraffe, hyaena, cheetah, impala, blue wildebeest, zebra, buffalo, kudu, waterbuck and more than 240 species of birds.

Climate: Hot summers (up to 40°C), with rain. Winters moderate.

When to go: Throughout the year.

How to get there: From Johannesburg take the N4 east to the Belfast turnoff (215 km), then follow the R540 from Belfast to Lydenburg, and on to the R36 to Ohrigstad (140 km). From Ohrigstad follow the R527 to Hoedspruit, and then to Motswari (154 km). M'bali lies 9 km from Motswari. Both have airstrips.

Opening time: 24 hours a day.

Getting around: Escorted game-viewing drives in the morning and evening; walking trails accompanied by rangers. There are also a number of beautiful hides for game-viewing.

Special precaution: Visitors are advised to take a course of anti-malaria pills before entering the park in summer.

Accommodation: Tanda Tula: There are seven 2-bedded rondavels. Motswari: Eleven 2-bedded and four single bungalows. M'bali: this camp has eight 2-bedded tents. There are conference facilities at Tanda Tula and Motswari. Ngala Game Lodge in the south has 20 two-bedded chalets and a luxurious safari suite.

Booking information: Tanda Tula: Safariplan, P O Box 4245, Randburg 2125 (telephone Johannesburg 8861810) or otherwise telephone the lodge through Hoedspruit 2322. Motswari and M'bali camps: Motswari Game Lodges, P O Box 67865, Bryanston 2021 (telephone Johannesburg 4631990/1). Ngala: Conservation Corporation, P O Box 1211, Sunninghill Park 2157 (telephone Johannesburg 8038421/8031810).

Tsavo National Parks – East and West

Location: Southeastern Kenya. Tsavo West lies 295 km from Nairobi and Tsavo East is 85 km further east. Map page 256, reference: I/J3.

Wildlife: Lion, leopard, cheetah, elephant, black rhino, giraffe, buffalo, impala, eland, kongoni, warthog, gerenuk, lesser kudu and a prolific birdlife.

Climate: Tropical, with high temperatures and humidity, and two rainy seasons – the 'long' rains falling between March and mid-June and the 'short' rains from November to early December.

When to go: Open all year.

How to get there: Follow the main Nairobi/Mombasa highway eastwards towards Mtito Andei – home of the Tsavo West headquarters and the main gate. Continue eastwards along the main road to Voi – be sure to follow the signposts to the Voi Gate – some 3 km from the town, and the main entrance to Tsavo East. The major lodges all have airstrips, and flights are available from Nairobi and Mombasa.

Opening time: 6 am to 6 pm.

Getting around: Most of the lodges and luxury tented camps offer guided safaris, and safari tours can also be arranged through operators in Nairobi and Mombasa.

Special precaution: Bring food.

Accommodation: Tsavo West has six or more camping sites, two well-equipped banda (hutted) camps at Kitani and Ngulia Safari camps, and two luxury lodges, Kilaguni Lodge (with an information centre) and Ngulia Lodge. Tsavo East has a campsite and wooden cottages at Voi Gate and the more sophisticated Voi Safari Lodge a few kilometres away. The exclusive Tsavo Safari Camp, reached via the Mtito Andei Gate further west, provides for the more well-to-do visitor.

Booking information: Book for Ngulia and Kitani Safari camps at Let's Go Travel, P O Box 60342, Nairobi, Kenya (telephone Nairobi 340331); or Abercrombie and Kent, P O Box 59749, Nairobi (telephone Nairobi 334955). For the Tsavo Safari Camp, contact Kilimanjaro Safari Club, P O Box 30139, Nairobi (telephone Nairobi 338888). Wildlife Conservation and Management Department, P O Box 40241, Langata Road, Nairobi (telephone Nairobi 501081).

Tsitsikamma National Park

Location: 200 km west of Port Elizabeth. Map page 256, reference: E14.

Wildlife: Bushpig, blue duiker, vervet monkey, caracal, honey badger, otter and abundant birdlife which includes the colourful Knysna lourie.

Climate: Rain throughout the year. Wettest months: May and October; driest months: June and July. Temperate climate.

When to go: The midsummer months between November and February are preferred.

How to get there: The turnoff to the park is about halfway between Knysna and Humansdorp on the N2, 9 km west of the Paul Sauer Bridge over Storms River.

Opening time: Open daily from 5.30 am to 9.30 pm.

Getting around: The major hiking trail is the 48 km Otter Trail which leads west, from Storms River Mouth. Other trails include the Mouth, the Loerie and the Blue Duiker. Suitably qualified swimmers, divers and underwater photographers can view the wonders of the sea on the Scuba Trail. There is also a snorkeling trail.

Special precautions: Strict regulations govern the lighting of fires. The Otter Trail demands forethought in equipping yourself, stamina (paths rise steeply to circumvent cliffs) and a head for heights.

Accommodation: At Storms River Mouth there are fully equipped two-bedroom beach cottages, a number of self-contained one-bedroom (four beds) cottages with kitchenette and bathroom and one- and two-bedroom oceanettes (beach apartments). There are camping and caravan sites. Washing and ironing facilities are provided, and there is a swimming pool. The De Vasselot section has camping and caravan sites, and the Otter Trail and Tsitsikamma Trail have overnight huts.

Booking information: The National Parks Board, P O Box 787, Pretoria 0001 (telephone Pretoria 3431991); or National Parks Board, P O Box 7400, Roggebaai 8012 (telephone Cape Town 222810).

Umfolozi Game Reserve

Location: About 270 km north of Durban. Map page 256, reference: G12.

Wildlife: Elephant, lion, black rhino, white rhino, blue wildebeest, leopard, giraffe and a variety of antelope and birds.

Climate: Hot, wet summers and balmy winters.

When to go: The reserve is open all year, but the best time to visit is during winter.

How to get there: Travel north from Durban on the N2. At 3,4 km north of the Mtubatuba turnoff, turn left to Nongoma. The Umfolozi turnoff is 27 km farther along this road.

Opening time: The gates open at sunrise and close at sunset (the times vary according to the time of year).

Getting around: Three-day wilderness trails, on which you hike and camp in the reserve's more remote areas, are the ideal way of seeing Umfolozi. A ranger accompanies each party (of no more than six people). Travel by private car, however, is the more usual method.

Special precautions: Take precautions against malaria. The water in the rest camp is chlorinated, but elsewhere it must be boiled or chlorinated as there have been instances of cholera in Zululand. Visitors should stay in their cars except at the designated viewpoints and game hides.

Accommodation: There are two hutted camps: Masinda and Mpila. These have fully equipped and serviced rest huts with separate kitchens and ablution blocks. There are also six chalets at Mpila and a nine-bedded lodge at Masinda. On a bank of the Black Umfolozi River lies the eight-bedded Sontuli Bush Camp. Downstream from Sontuli is Nselweni Bush Camp, which accommodates eight people in two-bedded units on stilts.

Booking information: To book wilderness trails and accommodation write to: The Reservations Officer, Natal Parks Board, P O Box 662, Pietermaritzburg 3200 (telephone Pietermaritzburg 471981).

Index

Photographic credits

Photographic credits for each page read from top to bottom, using the top of the picture as the reference point. Where the tops of two or more pictures are on the same level, credits read from left to right.

Abbreviations: ABPL – Anthony Bannister Photo Library; AI – African Images; PA – Photo Access; TCC – Trannies CC.

2 Anthony Bannister/ABPL
5 Peter Pickford/ABPL
6 Beverly Joubert/ABPL
7 Anthony Bannister/ABPL
9 Wayne Saunders/ABPL
10 Daryl Balfour/ABPL
11 J J Brooks/PA
12 Nigel Dennis/ABPL
13 Anthony Bannister/ABPL
14 Patrick Wagner/PA
15 Jeannie Mackinnon
16 David Steele/PA
17 Nigel Dennis/ABPL, Lex Hes
18 Daryl Balfour/ABPL
19 Lex Hes, Daryl Balfour
20 Anthony Bannister/ABPL
21 Anthony Bannister/ABPL
22 Joan Ryder/ABPL
23 Lex Hes, Carol Hughes/ABPL
24 Paul Funston/ABPL, Richard du Toit/ABPL
25 Anup Shah/ABPL
26 J J Brooks/PA
27 Lex Hes/PA, Clem Haagner/ABPL
28 Roger de la Harpe/ABPL, Anthony Bannister/ABPL
29 Wayne Saunders/ABPL
30 Robert Nunnington/ABPL, Daryl Balfour
31 Roger de la Harpe
32 Roger de la Harpe/ABPL
33 Clem Haagner/ABPL
34 Beverly Joubert/ABPL, Lex Hes
35 Lisa Trocchi/ABPL
36 Lorna Stanton/ABPL
37 Lex Hes
38 Peter Pickford/ABPL, Lex Hes
39 Lex Hes
40 Lex Hes
41 Richard du Toit/ABPL, James Marshall/AI
42 Lex Hes
43 Lex Hes
44 Lex Hes
45 Lex Hes
46 Daryl Balfour, Paul Funston/ABPL
47 Wayne Saunders/ABPL
48 C F Bartlett/PA
49 Both Anthony Bannister/ABPL
50 Both Daryl Balfour
51 Anthony Bannister/ABPL
52 Roger de la Harpe
53 Richard du Toit/ABPL
54 Anthony Bannister/ABPL
55 Lorna Stanton/ABPL
56 Terry Carew/PA, Roger de la Harpe/Natal Parks Board
57 Lorna Stanton/ABPL
58 Gavin Thomson/ABPL
59 Both Lex Hes
60 Nigel Dennis/ABPL
61 Duncan Butchart/AI, Roger de la Harpe
62 Roger de la Harpe/ABPL
63 Roger de la Harpe
64 Lex Hes
65 Rod Patterson/ABPL
66 Anthony Bannister/ABPL
67 Richard du Toit/ABPL
68 Wayne Saunders/ABPL
69 Anthony Bannister/ABPL
70 Duncan Butchart/AI
71 Clem Haagner/ABPL
72 Daryl Balfour
73 C F Bartlett/PA, Daryl Balfour
74 Anthony Bannister/ABPL
75 Nigel Dennis/ABPL, Daryl Balfour/ABPL
76 Daryl Balfour
77 Roger de la Harpe, Lex Hes/PA
78 Roger de la Harpe/ABPL
79 Anthony Bannister/ABPL
80 Daryl Balfour, James Marshall/AI
81 J J Brooks/PA, Clem Haagner/ABPL
82 Anthony Bannister/ABPL, Patrick Wagner/PA
83 Nigel Dennis/ABPL, Anthony Bannister/ABPL
84 Anthony Bannister/ABPL
85 Robert Nunnington/ABPL
86 Anthony Bannister/ABPL
87 Alan Binks/ABPL
88 Anthony Bannister/ABPL
89 Roger de la Harpe
90 Peter Steyn/PA
91 Lorna Stanton/ABPL
92 Dr Johan Kloppers
93 Roger de la Harpe/Natal Parks Board
94 Daryl Balfour
95 Anthony Bannister/ABPL
96 Roger de la Harpe/ABPL
97 Anthony Bannister/ABPL
98 Anthony Bannister/ABPL
99 Lex Hes/PA
100 Anthony Bannister/ABPL, Clem Haagner/ABPL
101 Lex Hes
102 Peter Steyn/PA
103 Lex Hes, Daryl Balfour
104 Colin Bell/AI
105 Roger de la Harpe/Natal Parks Board
106 J J Brooks/PA, Duncan Butchart/AI
107 Anthony Bannister/ABPL
108 Peter Pickford
109 David Steele/PA
110 Joan Ryder/ABPL
111 Daryl Balfour
112 Both Lex Hes
113 Lex Hes
114 Ken Gerhardt/PA
115 John Yeld/PA
116 Peter Pickford, Barrie Wilkins/PA
117 Nigel Dennis/ABPL
118 Anthony Bannister/ABPL, Dr Johan Kloppers
119 Lex Hes
120 Anthony Bannister/ABPL
121 Terry Carew/PA, Nigel Dennis/ABPL, Daryl Balfour
122 Anthony Bannister/ABPL
123 Clem Haagner/ABPL
124 Daryl Balfour/ABPL
125 Roger de la Harpe/Natal Parks Board
126 Clem Haagner/ABPL
127 Daryl Balfour, G L du Plessis/PA
128 Daryl Balfour
129 Daryl Balfour, Anthony Bannister/ABPL
130 Carol Hughes/ABPL
131 Anthony Bannister/ABPL
132 Lex Hes
133 Anthony Bannister/ABPL
134 Clem Haagner/ABPL
135 Lex Hes
136 Roger de la Harpe/Natal Parks Board, Richard du Toit/ABPL
137 Jean Laurie/PA
138 Koos Delport/PA
139 Peter Pickford
140 Anthony Bannister/ABPL
141 Clem Haagner/ABPL
142 Tim & June Liversedge/ABPL, Clem Haagner/ABPL
143 Rod Patterson/ABPL
144 Anthony Bannister/ABPL
145 Daryl Balfour
146 Daryl Balfour
147 Daryl Balfour
148 Clem Haagner/ABPL
149 Daryl Balfour
150 Anthony Bannister/ABPL
151 Robert Nunnington/ABPL
152 Clem Haagner/ABPL
153 Roger de la Harpe, Lex Hes
154 Clem Haagner/ABPL
155 Clem Haagner/ABPL, Daryl Balfour
156 Daryl Balfour
157 Daryl Balfour
158 Both Lex Hes
160 Clem Haagner/ABPL
161 Daryl Balfour/ABPL, Brenda Glyn
162 Lisa Trocchi/ABPL, Lex Hes
163 Lex Hes
164 Walter Knirr
165 Lex Hes
166 Anthony Bannister/ABPL, Clem Haagner/ABPL
167 Anthony Bannister/ABPL
168 Anthony Bannister/ABPL
169 Roger de la Harpe
170 Daryl Balfour, Daryl Balfour/ABPL
171 Anthony Bannister/ABPL
172 Anup Shah/ABPL
173 James Marshall/AI, Daryl Balfour
174 Wayne Saunders/ABPL
175 Lex Hes
176 Daryl Balfour
177 Daryl Balfour, Richard du Toit/ABPL
178 Daryl Balfour
179 Daryl Balfour
180 Lex Hes/PA, Herman Potgieter/TCC
181 Daryl Balfour
182 Richard du Toit/ABPL
183 Daryl Balfour, Anthony Bannister/ABPL
184 Lex Hes, Clem Haagner/ABPL
185 Clem Haagner/ABPL
186 Lex Hes
187 Both Lex Hes
188 Daryl Balfour
189 Anthony Bannister/ABPL, Daryl Balfour
190 Eric Reisinger/ABPL, Tim & June Liversedge/ABPL
191 Eric Reisinger/ABPL, Daryl Balfour
192 Daryl Balfour
193 Daryl Balfour
194 Lex Hes
195 Anthony Bannister/ABPL, Lex Hes
196 Roger de la Harpe/ABPL, Lex Hes
197 Lex Hes
198 Daryl Balfour, Anthony Bannister/ABPL
200 Lex Hes/PA
201 Terry Carew/PA, Dr Johan Kloppers
202 Ken Gerhardt/PA
203 Anthony Bannister/ABPL
204 Nigel Dennis/ABPL
205 Both Daryl Balfour
206 Nigel Dennis/ABPL, J J Brooks/PA
207 Peter Pickford
208 Richard du Toit/ABPL
209 G L du Plessis/PA, Daryl Balfour
210 Eric Reisinger/ABPL, Clem Haagner/ABPL
211 Anthony Bannister/ABPL
212 Both Lex Hes
213 Lex Hes
214 Anthony Bannister/ABPL
215 Anthony Bannister/ABPL
216 Robert Nunnington/ABPL, Roger de la Harpe
217 Shaen Adey/Natal Parks Board
218 Both Daryl Balfour
219 Daryl Balfour/ABPL, Daryl Balfour
220 Herman Potgieter/ABPL
221 Brandon Borgelt/TCC
222 Peter Pickford, Roger de la Harpe/ABPL
223 G L du Plessis/PA
224 Anthony Bannister/ABPL
225 Daphne Carew/ABPL
226 Tim & June Liversedge/ABPL, Nigel Dennis/ABPL
227 Nigel Dennis/ABPL
228 Anthony Bannister/ABPL, Daryl Balfour
229 Robert Nunnington/ABPL
230 Daryl Balfour, Anthony Bannister/ABPL
231 Anthony Bannister/ABPL
232 Both Lex Hes
233 Nigel Dennis/ABPL
234 Lex Hes
235 Rudi van Aarde/ABPL
236 Anthony Bannister/ABPL
237 Peter Pickford
238 David Steele/PA
239 Ken Gerhardt/PA
240 Anthony Bannister/ABPL, Jeannie Mackinnon
241 Philip Huebsch
242 Peter Pickford, Gavin Thomson/ABPL
243 Anthony Bannister/ABPL
244 Roger de la Harpe, Anthony Bannister/ABPL
245 Anthony Bannister/ABPL, Michael Meyer
246 Peter Pickford, Lex Hes
247 David Steele/PA
248 Anthony Bannister/ABPL
249 Daryl Balfour
250 Daryl Balfour, Jeannie Mackinnon
251 Anthony Bannister/ABPL
252 Daryl Balfour/ABPL, Richard du Toit/ABPL
253 Anthony Bannister/ABPL
254 Anthony Bannister/ABPL
255 Brandon Borgelt/TCC, Anthony Bannister/ABPL

Reproduction by Hirt & Carter Repro, Cape Town. Printing and binding by C & C Offset Printing Co, Hong Kong.

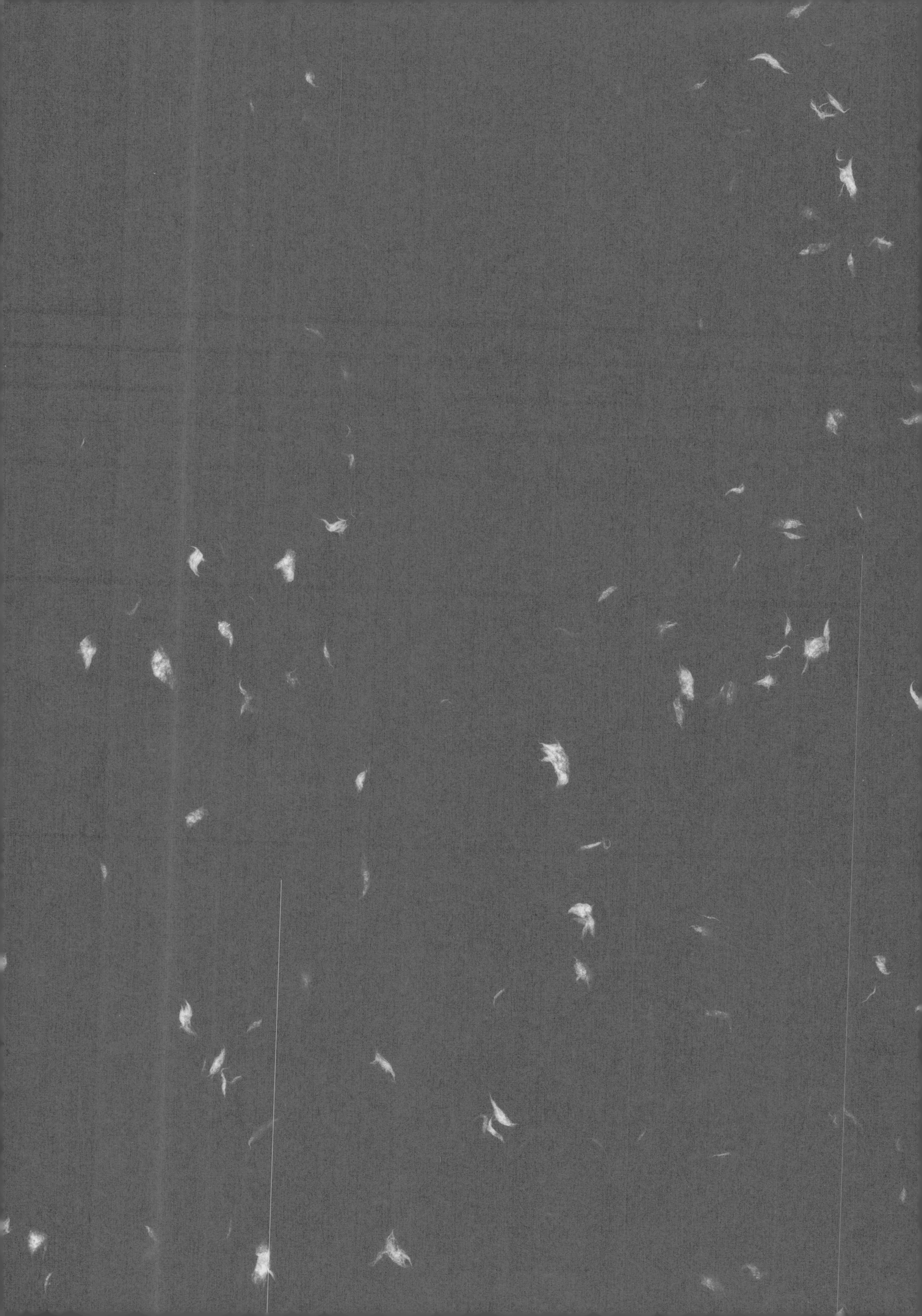